PREPARING OUR TEACHERS

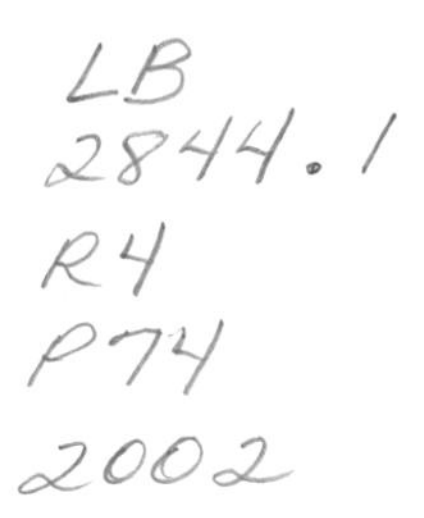

PREPARING OUR TEACHERS

Opportunities for Better Reading Instruction

The New Brunswick Group

Dorothy Strickland
Catherine Snow
Peg Griffin
M. Susan Burns
Peggy McNamara

JOSEPH HENRY PRESS
WASHINGTON, D.C.

Joseph Henry Press • 500 Fifth Street, N.W. • Washington, D.C. 20001

The Joseph Henry Press, an imprint of the National Academies Press, was created with the goal of making books on science, technology, and health more widely available to professionals and the public. Joseph Henry was one of the founders of the National Academy of Sciences and a leader in early American science.

Library of Congress Cataloging-in-Publication Data

Preparing our teachers : opportunities for better reading instruction / Dorothy Strickland ... [et al.] (the New Brunswick Group).
p. cm.
Includes bibliographical references and index.
ISBN 0-309-07445-2 (pbk. : alk. paper)
1. Reading teachers—Training of—United States. 2. Reading. I. Strickland, Dorothy S. II. New Brunswick Group.
LB2844.1.R4 P74 2002
428.4'071—dc21

2002151774

Printed in the United States of America

Contents

Introduction 1

Teachers of Young Children: The Agents and Champions of Literacy

Chapter 1 29

From Shopping Lists to Poetry: Forms and Functions of Written Language

Chapter 2 55

Making Meaning: Language Development and Comprehension

Chapter 3 81

Sounds, Letters, and Words: How Print Works

Chapter 4 119

Moving to Success: Motivating Children to Read

Contents

Chapter 2 55

Making Meaning: Language Development and Comprehension

Chapter 3 81

Sounds, Letters, and Words: How Print Works

Chapter 4 119

Moving to Success: Motivating Children to Read

Introduction

Teachers of Young Children

The Agents and Champions of Literacy

A class of young children sits around her—each with a different past and unknown potential, each with a unique yearning. Some are already beginning to read a few words or even sentences. For others the print in books is still a confusing jumble of letters—or merely squiggles—upon the page. From these varied beginnings the teacher's goal is to give each child something he or she needs, something that changes a life: the ability to read and write.

During the gray days of February, the teacher is a coach who can inspire and motivate the most discouraged student. Sometimes she seems to be a mind reader, too, who sees inside children's heads to figure out how they are progressing, when they get stuck, and what kind of practice or explanation they need to succeed. She gets each child to ask himself or herself the important questions that tease out meaning from the text.

At times she may be the fearless guide, ever ready to take her class on an illuminating detour. One day it's the crazy fact that *cave* and *have* sound altogether different. She helps them remember about *save* and *gave,* delving into the rhyme and a reason, the regular pattern and the exception.

Another time she gets them into the long and short of it—this author's *"long hour"* is no longer than the standard 60 minutes, and that other book's *"short two weeks"* last a full 14 days. How can sense be made of that? The teacher reminds the class that last week at the zoo they were sure her watch was wrong; was it really 11:30 already? That seemed to be a short 3 hours, they agree, because they didn't want it to end. They see that a writer chooses a word to show a feeling in a subtle way.

They learn to read not only between the lines through their background knowledge but also within, under, and around the words.

All along the way she is caring and sensitive to individual needs while able to maintain order in the classroom. She can build on the many cultural resources in her class, enthusiastic to learn about cultures and languages new to her. She discovers which boy or girl loves science, sports, cooking, or animals. She feeds them all sorts of reading matter—books, recipes, magazines, science experiments, and sports pages to incite curiosity, even passion, in the context of reading and print.

A master teacher is at times a consummate technician who relishes the chance to work on the nuts and bolts of written English, the expected sounds and the exceptional ones, the possible pitfalls and quirks—consonant digraphs, silent vowels, prefixes, suffixes, homophones, and homographs. At a moment's notice she transforms into a dramatist who brings color and life to stories, plays, and poems—the literal, the implied, the metaphorical, the tone, the style, and the detail. Another quick switch and she is a brilliant conversationalist, drawing out even the shiest children and thus stretching their language skills and background knowledge.

To watch a master teacher in action is like watching an artist immersed in her discipline, drawing on an array of techniques, skills, and visions of beauty to create distinct pictures with each boy and girl.

THE HIGH SCHOOL CLASS OF 2011 BEGINS TO READ

A long-term study is following the progress of 20,000 children who started kindergarten in 1998. The study's current results show the enormous variety related to reading that teachers handle day by day:

- Thirty-seven percent of entering kindergarten children know that print reads left to right, know how to sweep back to the left when a line of print ends, and know where a story ends on a page. Others have less familiarity with print; 18 percent show none of this knowledge.

- Sixty-seven percent of entering kindergarten children recognize the letters of the alphabet; 33 percent cannot do so; 31 percent of children connect letters to sounds at the beginning of words, and 18 percent do it at the ends of words as well. The letter-sound connection is still a mystery for about half of the children.

- As they are about to enter the first grade, 74 percent of children connect letters and sounds at the beginning of words; 54 percent make the connection at the ends of words as well. Only 14 percent recognize common words on sight, and only 4 percent can understand simple passages when they read on their own.

- Leaving first grade, almost all children connect letters to sounds at the beginnings (98 percent) and ends (94 percent) of words; 17 percent are not yet proficient at recognizing common words on sight. Half the children understand simple passages, but the others still need work in the second grade to read such passages on their own.

- Some differences in proficiency are associated with children's gender, age, race, ethnicity, family economic status, family type, and mother's education. But the highest quartile for reading proficiency includes children from all the different backgrounds.

SOURCES: http://www.nces.ed.gov/ecls
West, J., & Denton, K. 2002. *Children's Reading and Mathematics Achievement in Kindergarten and First Grade.* Washington, DC: National Center for Education Statistics.
West, J., Denton, K., & Germino Hausken, E. 2000. *America's Kindergartners.* Washington, DC: National Center for Education Statistics.
West, J., Denton, K., & Reaney, L. 2001. *The Kindergarten Year.* Washington, DC: National Center for Education Statistics.

This sketch barely touches the surface of all that a master teacher can do—from the annual, monthly, and daily plans, accountable to hundreds of "must-teach" skills, to the thousands of minute-to-minute actions, choreographed on the spot for the unpredictable needs of young and vulnerable children. By year's end she has helped to create competent readers, as if wresting the spirit from the stone—some more advanced than others, but all most definitely on their way to literacy.

THE ESSENTIAL PROFESSION

- Eighty-nine percent of Americans say it is very important to have a well-qualified teacher in every classroom.
- Eighty-percent agree strongly that fully qualified teachers should be provided to all children, even if that means spending more money.
- Seventy-seven percent say it is a high national priority to develop the professional skills and knowledge of teachers throughout their careers.
- More of the public (60 percent) identifies investment in teachers as the most crucial of strategies for improving student learning, topping other strategies like setting academic standards and instituting testing programs.

SOURCE: Haselkorn, D., & Harris, L. 2001. *The Essential Profession: American Education at the Crossroads.* Belmont, MA: Recruiting New Teachers.

No wonder we are sometimes tempted to believe that great teachers are born, not made. So great is the challenge. So important the job. So immense the bureaucratic obstacles.

"It all comes down to the teacher," parents are notorious for saying—and for competing to get their own children into certain classes taught by the current stars of the school. It is hard to argue with them. Nothing in this world can replace the power of a great classroom teacher during a child's formative years—not fancy computer labs, or great libraries, or after-school enrichment programs. Teaching, even in this technical and complex age, remains essentially a human operation. Every parent wants the best for each child. We need more quality to go around. We need more quality teachers to stay around.

Too often, though, effective teachers are enticed away from the classroom. Traditionally, the only way to advance on a teaching career path has been to become a guidance counselor, principal, curriculum supervisor, or superintendent. The career trajectory should celebrate greater mastery of and responsibility for classroom teaching, rather than movement out of the classroom. Such a trajectory, supported with excellent professional development, would provide momentum for quality classroom teaching.

If they are to excel in their craft and be satisfied with their careers, even the most talented and gifted individuals require a good foundation and apprenticeship, a lifetime of challenging work environments, and ongoing education. Some great teachers may have been born with a special talent, but even they need professional preparation to develop their potential. And many more great teachers forge their skills step by step, learning about subject matter, about how to teach, about how to observe and assess children learning, and about how to improve their own instructional practices. They learn to teach through their entire working lives.

Most important, teachers learn how to get children to become good readers. They build on a firm foundation of language development and integrate three aspects of skilled reading: identifying printed words, constructing meaning, and developing fluency.

Professional Development *Is* the Job

Anthony Alvarado, Chancellor for Instruction, San Diego (California) City Schools

It's a big mistake to think that teaching is what we do every day and professional development is an occasional seminar or workshop or institute. No! The job is professional development, and professional development is the job.

SOURCE: American Federation of Teachers/National Education Association Conference on Teacher Quality, September 25- 27, 1998, Washington, DC.

Yolene Medard, Classroom Teacher, Grade 2, New York, N.Y.

We have staff developers in our school. The most powerful have been the literacy developers. All the second-grade teachers meet weekly with Lisa. What's great is that she comes into my classroom every week on Tuesday afternoon. Sometimes she gets up and models minilessons for me to learn from, and she watches me teach, too. I sit in with her while she conferences with a child and make note of what goes on, or sometimes I do the talking and she listens. That has been very effective. I can count on her. I know she's in there 45 minutes once a week. The children develop a relationship with her. I have developed a relationship with her. When we sit down to meet, she gives me very specific feedback on my teaching strategies and approach. She is very positive and constructive—even when she's telling me about things I need to change. Then she builds on what I've done and asks, "So what are you going to do next?"

Formal classes help, too. Four years ago I went back to school to get a reading certificate. Even though I received a great education as an undergraduate, there were just some things I needed to know more about. I was teaching children who had few, if any books at home, some whose parents didn't graduate from high school, some whose parents didn't speak English. Some had no one to help with homework; they were basically on their own. When I began teaching 10 years ago, I would never have led a group that just went over the "th" sounds or "wh" or "sh." But now I do that with just the six kids who need it, not the other 20 who understand it. I'm very very explicit for some kids. I let them know we're just going to work on this sound until they have it.

Last year I went for certification through the National Board of Professional Teaching Standards. It taught me a lot. It was a year of looking critically at what I did in the classroom and changing to reach more of the kids more of the time.

To provide the best instruction for children learning to read, there is no question that teachers should be provided with far more and far better preservice and in-service education. Universities, colleges of education, schools, school districts, state education agencies, politicians, parents, teachers, and researchers—everyone must

SUPPORTING SKILLED READING

pull together to meet the challenge of better preparing our teachers, not only at the beginning of their careers but throughout their professional lives.

This book is about that challenge. It is also about our firm belief that teachers are the agents and the champions of literacy.

how this book came to be

We, the authors, are committed to improving children's reading instruction. This book follows up on the findings of a landmark study undertaken by some of the nation's leading reading researchers.[1] Four out of five of us were part of that project, which was carried out under the auspices of the National Academy of Sciences from 1995 to 1999. Our goal then was ambitious. We reviewed current

[1]See the Notes at the end of this chapter for those involved with the original committee and those who assisted with this book.

research to answer questions about which children are at risk of reading difficulties; what the process of skilled reading looks like; and which individual, family, social, school, and classroom factors are related to success in reading. We examined the methods and results of prominent approaches to reading instruction and intervention for reading difficulties. And we met with teachers, principals, and remediation specialists from all over the country.

In the end we offered a plan for how the nation could prevent reading difficulties, emphasizing that *good classroom instruction is the primary prevention.* The group discovered considerable consensus and concluded that there was little life left in the infamous "Reading Wars"—the contentious and overly simplistic ideological struggles well known nowadays to the public in the form of "whole language versus phonics." Our conclusions were published by the National Academy Press in an academic report called *Preventing Reading Difficulties in Young Children.*

The study was widely covered in the news media. The *New York Times, Boston Globe, Los Angeles Times,* and the *Today Show* all reported on the study's findings. *Newsday* described it as "the largest study of its kind ever attempted, which tackles some of the most explosive issues in education." Spanish and Chinese editions of the report were published. We were pleased that people responded to our call for a careful, complex, and integrated approach to early reading instruction rather than a single magic bullet or an eclectic, compromised mélange.

Because so much was at stake, we strove to reach a larger audience of nonacademics who impact children's lives each and every day. The result was *Starting Out Right: A Guide to Promoting Children's Reading Success,* which included techniques and activities for both the classroom and the home, from birth to third grade. We hoped to reach a fair number of parents, in-service teachers, day care providers, school board members, and policymakers by relating the research results directly to practical information.

To our delight and surprise, *Starting Out Right* has sold more than 150,000 copies. The success of these two books has motivated us to go a step further and now speak directly about teacher preparation. Arguably, this is our most important project yet. Research on education is not worth much if teachers do not have the support they need to put it to work in classrooms.

In this project we have been joined by a faculty member from Bank Street College of Education—one of the nation's most prestigious teacher preparation institutions. We call ourselves the New Brunswick Group because it was in New Brunswick, New Jersey, that we came together to begin brainstorming, sharing our ideas, our research, and our enthusiasm for this project. Our collaboration has been supported by a prominent membership association of educators and a private foundation heavily committed to improving education.

We knew that, in the years since our earlier books were published, reading in the early grades had been the topic of a great deal of state and federal policy initiatives and many widely distributed publications. Acrimony again arose about how to teach reading. The focal point for controversy has been phonics—patterns of links between letters and sounds—and phonemic awareness—knowledge about the sounds within spoken words.

Our earlier books take a clear position on the relationship among the different aspects of reading and we still stand behind it: According to the best-available knowledge, all children should have the opportunity for high-quality instruction that integrates printed word identification, meaning, and fluency. That is what prevents reading difficulties, and that is what starts children out right.

Too often, though, the message gets lost in the move to legislative mandates, executive initiatives, published curricula for teachers or children, or interpretations of research. Sometimes policymakers or writers overemphasize one part of the complicated reading process, claiming it is a redress for past neglect. Other times the emphasis is not intended or may be merely in the eyes of a partisan beholder. However originated, the loss of an integrated whole creates a vacuum that more controversy about reading instruction rushes in to fill.

Worry about time is the fuel for these controversies. Do we have time for instruction in phonemic awareness and phonics, which can improve word identification, without giving up the other aspects of skilled reading? If we agree to work on those topics, will we have too little time for other matters that children also need?

Happily, we can report that good news has been published since our earlier books came out. The news comes from the National Reading Panel, mandated by Congress to examine the state of research about early reading instruction. The panel undertook a metanalysis to examine differences among existing studies and to see if the results are consistent enough to guide practice and policy.

Overall, the panel reports that phonemic awareness and phonics instruction show consistently positive, small-to-moderate effects on children's word-reading outcomes. The analysis also shows that spending more and more time on phonemic awareness and phonics is *not* necessarily better. In fact, the best effects from phonemic awareness instruction were found with programs that used between 5 and 18 hours total in the course of a school year—a pittance if we remember that the number of school hours per year is about 1,080! Nor do phonemic awareness and phonics instruction need to be dragged out or done over and over throughout the elementary grades. It is time in kindergarten and first grade that has the real payoff. We need not consume time essential for other aspects of reading instruction—and all the other important matters that children meet in the early years of schooling. (See the Resources at end of this chapter for references to basic reviews of learning to read, including the National Reading Panel report.)

As we began this project, we knew, too, that there were requirements and suggestions, legislated standards, and detailed proposals about what one should know and be able to do in order to teach reading in the early grades. Various lists come from agencies, organizations, and associations in the government and private sector. Most begin with what children need to know and do and backtrack from that to what teachers need to learn; many have useful technical details. Some change from time to time, and new ones crop up. (See the Resources at end of this chapter for sample references about standards for teachers.)

Despite the considerable amount of activity, there are no clear answers about what needs to be done in teacher education and professional development. In an ideal world, studies would examine the path from teacher education programs through teacher learning to improved school practice to student learning and progress. That is a long and windy road and one that is difficult to study. Reviews of the literature attest to the vigor of inquiry about teacher education and professional development in general, and a few studies directly address teacher preparation regarding the teaching of reading in the early grades. (See the Resources at end of this chapter for sample references about teacher education and professional development.)

It seems as if disputes outnumber solutions when it comes to the content, contexts, and processes in teacher education programs and professional development plans. When, if ever, should teachers, as learners, "sit and get" information, "make and take" classroom materials, "see, try, and tinker" with the practices of master teachers, "reflect and react" on student data and research findings? Should teachers of teachers be lecturers, coaches, facilitators, mentors, peer groups, or all of the above? Will the technologies of distance learning and video- and computer-based cases fit into productive programs? Is it correct to think of teacher learning as an individual matter or is it best thought of as a coordinated school or district matter? Can someone teach and learn to teach at the same time? Can teacher education and professional development be vehicles for school and curricular reform? Can new curriculum materials and district or state mandates have a positive impact on teacher education programs and professional development plans? What is the relationship between teacher education and professional development on the one hand and retention and mobility of the work force on the other? Although sometimes the rhetorical tides run high, current evidence and argument are too fragmented and scanty to settle the issues.

Our contribution is about the *content* of teacher preparation for teaching early reading. It is the common ground that can be interpreted in a variety of contexts for teacher learning. For effective action to improve teaching and for research on teacher education and professional development, such common ground is needed. The content we call for can be addressed with whatever modes prove effective for teaching adults so that they are able and disposed to use what they learn in their daily work life.

Our framework is not based on the components coming out of a task analysis of individual skilled reading. Nor is it a refinement of the curricula and syllabi of teacher certification programs, although the five elements in it can be related to traditional disciplines in such programs (see, for example, pp. 285-287 in *Preventing Reading Difficulties*).

We looked instead for a perspective with a close tie between children's achievement and teacher education and professional development. Make no mistake; we understand that the effectiveness of teacher education and professional development is mediated by, sometimes overwhelmed by, practices in schools. Still, we wanted to think about what teachers need to know and be disposed to do at the same time that we thought about why it matters for children—a perspective that would integrate the "what" and the "why" of the content of teacher education and professional development.

Recently, a case has been made for "opportunity-to-learn" standards that focus on whether schools are providing the conditions children need to succeed (see Darling-Hammond, 2000a). We have such a framework from our work on *Preventing Reading Difficulties*. There we summarized what children need in order to become accomplished readers:

- opportunities to become familiar with the forms and uses of written language;
- opportunities to develop the language and metacognitive skills required for reading comprehension success at every stage of literacy development;
- opportunities to grasp how the words of the language are structured and are represented in print by the letters of the alphabet;
- opportunities to become enthusiastic about learning to read and write;
- opportunities, if at risk of reading failure, to be noticed early enough and to be offered enriched experiences and/or intensified instruction in school.

In *Preventing Reading Difficulties* we list six opportunities that should be provided for children, separating early prevention and early intervention. Because this book focuses on classroom teachers rather than the full range of school and community personnel involved with reading difficulties, we collapse those two and organize our thoughts around the resulting five major opportunities.

These opportunities are oriented to activities that teachers and schools should engage in. We contend that a good classroom teacher of reading learns to provide these five opportunities—better and better, more and more often, for each child. Teacher education and professional development can be judged by the extent to which both help teachers provide children with these crucial opportunities for becoming full members of our literate society.

We rely on these five major opportunities as the backbone of Chapters 1 through 5 of this book. In each chapter we discuss one aspect of what should be provided for children learning to read and offer concrete information about what teachers, in turn, must know and be disposed to do in order to provide that opportunity for each child in the classroom. To help demonstrate our vision, we supply several vignettes that dramatize prepared teachers in action. These are composites, based on our many collective years working with teachers in the field. In addition, sprinkled throughout the book are quotes from real-life teachers, masters and novices, who told us about their journeys toward better teaching.

To make our vision more specific, we provide samples of activities that teachers and teacher educators can use—in-service and preservice. Some activities are about teaching lessons or parts of lessons—planning, developing, observing, simulating, practicing, reflecting, and repairing when needed. Other activities are more about teacher learning—ways to develop understanding of concepts or practices. Some activities put teachers and teachers-to-be in the learner's shoes—tasks designed to remind them of the complexity of (and complex emotions associated with) different aspects of learning to read.

In this book each chapter has a list of resources at the end to substantiate the issues raised in the chapter and to provide starting points to elaborate the topics beyond what can be covered here.

To make our vision complete, we need more. We need to instigate action. Policymakers, educators, school administrators, and the public need to provide the crucial opportunity *for teachers*—the opportunity for them to improve their practice. Teachers need resources, reasons, and protective settings so that they can see and use and test and improve, so that they can continuously reflect on what is working and what more is needed to teach children to read.

what's at stake

A devastatingly large number of people in America cannot read as well as they need to for their own success and for the public good. Large numbers of school-age children, including children from all social classes, face significant difficulties in learning to read, but problems are especially likely among poor children, among children who are members of racial minority groups, and among those whose native language is not English. This is of grave concern to the entire nation. According to U.S. census data, nearly 40 percent of the nation's children currently belong to racial minority groups. Seventeen percent live in poverty.

The National Assessment of Educational Progress—known as the nation's report card—has found that two-thirds of all fourth graders tested fall below the level the federal government considers acceptable. Most disturbing is that despite decade-long efforts the test found a widening gap between the very best students and the very worst.

We must concern ourselves with the children in this country who do not read well enough to meet the demands of an increasingly complex world. To contribute to society in the twenty-first century, today's children will need to reach a high literacy threshold. They will have to read far more challenging material than students of yesteryear. They will use printed matter to solve problems independently—even in the entry ranks of the work force. They need and deserve schools and classrooms

FOURTH-GRADE READING IN 2000—THE NATION'S REPORT CARD

Only a third of the nation's fourth graders read proficiently according to the 2000 National Assessment of Educational Progress:

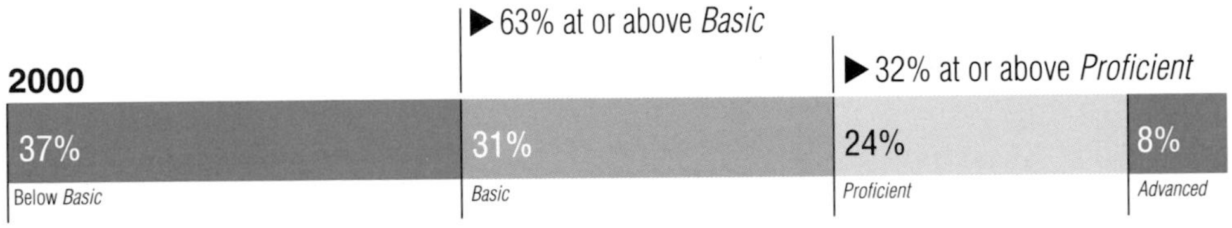

2000	▶ 63% at or above *Basic*	▶ 32% at or above *Proficient*	
37%	31%	24%	8%
Below *Basic*	*Basic*	*Proficient*	*Advanced*

What is proficient reading in grade 4?

"For example, when reading literary text, Proficient-level fourth graders should be able to summarize the story, draw conclusions about the characters or plot, and recognize relationships such as cause and effect. When reading informational text, Proficient-level students should be able to summarize the information and identify the author's intent or purpose. They should be able to draw reasonable conclusions from the text, recognize relationships such as cause and effect or similarities and differences, and identify the meaning of the selection's key concepts."

SOURCE: U.S. Department of Education. Office of Educational Research and Improvement. National Center for Education Statistics. 2001. *The Nation's Report Card: Fourth-Grade Reading 2000*, by P. L. Donahue, R. J. Finnegan, A. D. Lutkus, N. L. Allen, & J. R. Campbell. Washington, DC: Author.

where 100 percent of the children are literate and deep levels of literacy are the norm. They need teachers who know how to provide them with the opportunity to learn to read proficiently.

what are teacher education and professional development?

There has been some concern about whether a teaching career attracts less able college students, but a recent study shows that teachers are as able as those who go into, for example, law, medicine, engineering, and accounting. Applicants for teacher education should be expected to have a good liberal arts and sciences education. Some basic courses in the behavioral, cognitive, and social sciences are crucial for preeducation students, just as biology and chemistry courses are for premed students.

Schools, departments, or programs of teacher education are responsible for specialized courses. Across the country, teacher education requirements vary

HOW TEACHERS COMPARE

As a group, teachers score relatively high in literacy.

- About half of teachers score at the two highest levels, while only 20 percent of other adults nationwide do so.

- On average, teachers perform as well as other college-educated adults. Teachers who take graduate studies or degrees match the performance of other adults with graduate study experience.

- In prose literacy, teachers score higher, on average, than managers and administrators, real estate and food service managers, and designers. They perform at a similar level with lawyers, electrical engineers, accountants and auditors, marketing professionals, financial managers, physicians, personnel and training professionals, social workers, and education administrators and counselors.

SOURCE: Bruschi, B. A., & Coley, R. J. 1999. *The Prose, Document, and Quantitative Skills of America's Teachers.* Princeton, NJ: Educational Testing Service.

considerably. In some states, teacher credentials are available after a four-year bachelor's degree that includes a set of prescribed courses in an accredited college or university. In other states, teachers must have an additional year, some with, some without, a master's degree. Provisional certification may be a first step, with permanent or recertification requirements that include passing additional formal courses and/or assessments of knowledge and classroom practice. Some states use formal sit-down tests as part of the credentialing process.

Whatever the technical and legal requirements, a classroom teacher's preparation should be viewed along a continuum—a lifelong journey of professional development that does not end with a teaching degree or state certification.

State and local education agencies or individual schools are responsible for professional development for teachers. Professional development may take the form of a mandatory annual workshop or an optional summer institute. It may include seminars that a principal arranges to study findings from newly published reading research or to introduce a new resource for children's literature. It may be a series of meetings to help teachers learn about the literacy practices of an ethnic group recently arrived in the school cachement area. Or it may be regularly scheduled school-site sessions in which colleagues work to coordinate curriculum objectives across grade levels. Mentoring for novice teachers is another part of a professional development program. Professional development means staying up to date with current research and may include partnerships for contributing to the research base. For part of their continuing education, individual teachers may enroll in night or summer courses or advanced degree programs at a college or university.

Valued professional development sessions are not costly lectures by outside experts or the frequently provided "off-the-shelf" training from publishers of children's textbooks. For the report *Teacher Quality: A Report on the Preparation and Qualifications of Public School Teachers,* the National Center for Education Statistics asked teachers about the relative values of different professional development activities. They credited some activities with improving teaching a lot. The activities

PHASES IN A TEACHING CAREER

Clinical: Still enrolled in formal teacher education, novices observe classes and may tutor in a supervised program as part of their course work. At the apex of the clinical phase is the student teacher, supervised by an expert classroom teacher. In most states, teachers are eligible for entry-level certification after they perform well in this on-the-job experience.

Intern: These teachers are in their first year or two as the paid full-time head teacher of a class. Interns sometimes participate in school district-sponsored induction programs, which may lead to a higher-level teaching certificate or a master's degree.

Residency: At this stage, a teacher is a developing professional in his or her third through seventh year of teaching. He or she has limited but growing responsibility to mentor student teachers in their clinical phase.

Mentor: Expert teachers are able to function as instructional leaders, coaches, and/or organizers of professional development in a school or district. The essential feature of a mentor teacher is more than some arbitrary number of years of experience. The essential feature is teaching expertise, recognizable by peers, attested in practice and student outcomes, and used to guide and support teachers in the clinical and intern phases.

that rose to the top were team planning periods, mentoring by another teacher, and regular collaboration with others.

There is a good deal of current research on the forms and processes of teacher learning to assess which activities have the best results for both teacher learning and children's achievement. Our expertise is about the content rather than the form or process of teacher education and professional development. We expect that much of the content can be done well (or poorly) in a variety of ways. The activities we provide are samples, examples that we and our colleagues have found useful in our teaching of teachers.

The studies of teacher learning are beginning to tease apart the value of the different roles of those available to help teachers learn—professor, mentor, coach, facilitator, staff developer in the school or from an outside source, assistant principal for instruction, curriculum supervisor. We believe that people filling any of these roles and working on the teaching of reading need a firm grounding in the content we describe and the disciplines behind it. Activities like the ones we provide require depth and breadth of knowledge by those teaching teachers.

An especially important issue about the processes of teacher learning is the relationship between the more abstract presentations and the more practical work—the

PROFESSIONAL DEVELOPMENT CAN TAKE MANY FORMS

The trick is to have close ties to the specific problems that participants face in their classrooms and to have connections to the wider world of policy and research. The school or district may provide a facilitator to organize and sustain efforts like the following:

Group study: Teachers meet in grade-level teams once a week to discuss solutions they have tried; relevant readings; or input from mentors, coaches, or supervisors. The same topic can span weeks, even months, of the group study. A topic like estimated or invented spelling may lead to a study of articles about phonemic awareness and decoding.

Case study and chat rooms: Meeting over the Internet, a teachers' group studies practice problems or cases that challenge the problem-solving abilities of groups. They learn to identify and define teaching and learning problems in situations that approach the complexities of actual classrooms. They evaluate alternate solutions from numerous angles. After considering prepared cases, individuals "case" their own classrooms and focus the group effort for maximum utility. A topic like comprehension may narrow to focus on how to teach children to use summarizing and prediction strategies together.

Focused observations and action research: Pairs or small groups of teachers collaborate with a researcher to identify a question about instructional practice. They develop a system to document observations in their classrooms. They observe each other's classrooms and then examine the data and develop plans for improved practice that call for a new round of observations. Starting with a question about fluctuations in fluency during oral reading, they may try varying the kinds of materials to increase successful practice.

Mentoring: A more experienced teacher works with one or more novices. Mentors teach model lessons and follow up with notes about the plan and the happenstance, the success, and how to repair the parts that went awry. Mentors observe novices' regular classroom practice and provide detailed feedback. Mentors consult about instruction for specific students. Mentors get coaching and information through ties with district personnel and teacher educators. Mentors say they learn as they teach teachers.

practicum, laboratory sections, classroom observation, tutoring, clinical experience, student teaching or internship, classroom intervisitation, professional development schools, coaching. More abstract material in lectures, discussions, and readings gets a true comprehension test when put into action with children. Reciprocally, actions with children can mount as teacher expertise if they are subject to public discussion and interpretation in the light of theory and evaluation of outcomes. Teacher preparation that uses teaching cases provides practice, vocabulary, and routines that teachers can apply to learn deeply from the "case" of their own practical experience. The important question is how and when to best manage integration of the abstract and the practical. (For more discussion, see Clandinin and Connelly, 1996, and Hoffman and Pearson, 2000, listed in the Resources.)

a firm footing for learning to teach reading in the early grades

Because reading touches all content areas—from sciences and social studies to literature and philosophy—good teachers benefit from being well read themselves and knowledgeable in many disciplines. Information and values, procedures and approaches, dispositions and inspiration—the yield of a good education—are what effective primary grade teachers rely on every day. We are concerned that even before teachers-to-be declare their education majors in college, they may not be getting the sort of foundation they need.

As part of a solid liberal arts and sciences education, we believe it is useful for future teachers to study a language that is not their native one. Like any student, they can enjoy new worlds in the literature and culture that a new language presents to them. Most important, as teachers-to-be grapple with learning to speak, read, and write in a new language, they can develop empathy for students they will meet for whom English is a new language. Foreign-language study is not currently a universal requirement in liberal arts and science programs, but we think it is a valuable part of a well-rounded education and is especially useful for those preparing to teach a diverse student body.

Education courses should rest on a solid bedrock of knowledge. In an education program, department, or school, there are bodies of knowledge and skill that transcend our specific concern with teaching reading in the early grades. We believe educators of reading teachers should be able to take the more general knowledge, skills, and dispositions for granted. Until the preeducation and education courses are operating in this sort of ideal world, though, preservice curricula and professional development plans must take the time to develop the resources to augment several kinds of fundamental knowledge. Such foundations support good teaching practice in general. While they are not specific to preparation and development for effective teaching of reading, they are crucial for it.

Knowledge in the Behavioral and Cognitive Sciences

Those who want to teach young children in preschool and the primary years need to learn how reading, writing, and language develop in young children. Preschoolers not only talk to express meaning but also use symbols, drawings, pretend stories, and many forms of play. These activities provide starting points for reading. Teacher candidates should take at least one course in child development that includes language and literacy development.

In their daily lives, early childhood educators will grapple with many questions about what makes a person do one thing and not another or find some things easy and others hard. They will want to know how children think and approach problems, what leads to misunderstandings, what memory is and how it works. They will wonder at what point youngsters are ready to "pay attention." They will constantly ask themselves how they can motivate children to achieve and to pursue their own curiosities in books. We strongly recommend that early childhood educators take at least an introductory course in psychology to help prepare themselves for these challenges.

Knowledge in the Social Sciences

More than ever, classroom teachers will want to be worldly. As immigration and demographic shifts continue, teachers will encounter greater numbers of children who come from non-English-speaking families. For this reason they'll want to be prepared to treat home-language and culture differences as the assets they can be, not just for the individual child but for everyone. They should learn about the social and cultural issues relevant to the children in their regions. They will want to learn how different populations use writing and speaking, how education is viewed, how literature is used and shared by other cultures. At least one course in local sociology, anthropology, or regional American studies is a must.

Knowledge of Language and Literature

Elementary school teachers should be lovers (or at least admirers) of language itself, not only books. In addition to the literature and rhetoric of a good liberal arts education, we recommend that all preservice teachers take a basic course on the structure, history, and variability of the English language. This is currently not very common. Yet a kindergarten teacher should know the difference between phonemes and phonics. A first-grade teacher must be able to tell a dialect pronunciation from a mistake in reading. A fourth-grade teacher should understand the value of pointing out cognates—words that are similar in two languages—to the children who speak both English and Spanish. Only teachers who understand the nuts and bolts of the English language—oral and written components—will be able to get inside children's heads to find out where a student is going wrong and to provide the right help when needed.

Every teacher of young children should be wise in the ways of excellent children's literature. The problems, the triumphs, the silliness, and the depth—this body of work resonates with children's lives. Good authors have always served society in this way; their work tells us about ourselves and each other. Teachers need to have ways to connect with children emotionally and intellectually. Sharing good children's literature is both a place to start and a culmination. Teachers need to know what is out there and what makes some of it so good.

Knowledge of Assessment, Management, and Family Involvement

As part of foundation knowledge, teachers must be prepared to understand, choose, use, and interpret a variety of assessment tools. They need techniques to screen for potential problems and determine individual strengths and weaknesses. With these skills they can tailor instruction according to individual needs and evaluate when to stay the instructional course and when to begin searching for alternatives.

They will also want to learn how to structure the learning environment using a wide range of materials, technologies, and intervention programs. They should be able to run emotionally, physically, and socially safe classrooms. Good teachers should also be taught the best techniques for coordinating with families. They will want to know how to draw on school and community resources, making referrals and coordinating with follow-up services.

Knowledge of Standards

A final part of the foundation of good teaching is another that is seldom adequately focused on in teacher education. Course time needs to be spent on standards—the state or district mandates about the topics and skills to be covered in school subjects. Some states tell classroom teachers exactly what to teach every 6 weeks. Others simply tell them what must be learned by the end of each year or every several years. Some distinguish reading skills from writing and spelling; others lump reading into a broad array of language arts skills. Standards are often tied to student testing and assessments, often with high political and perhaps personal stakes attached to them.

Teachers need to know where the standards are to be found, what they mean, and how they apply, and teachers need to solve problems about teaching in the context of standards. To be well prepared to teach reading, educators need to know how well those standards align with the reading programs they are using and the tests their students will take. They will want to take the standards into account as they are making decisions about curriculum, activities, and materials and as they are reviewing their successes or failures for a given year. We cannot leave it to novice practitioners to make the bridge between the ideas about reading in education courses and the mandates about reading in the law.

With these foundations in place, aspiring teachers are ready to profit from courses and professional development particularly dedicated to preparing them to teach reading in the primary grades. While many will arrive for pre- and in-service education without some of this firm footing, we must go for the gold and help aspiring teachers build the rest as we go along.

who this book is for

This book is for parents and other citizens who need a better understanding of what underlies good reading instruction by classroom teachers. It is a basis for conversations between teachers as learners and those providing resources for their learning. It can be a touchstone to examine where the needs are in a particular school, grade, or individual.

It is also for the people who prepare teachers-to-be during their undergraduate and graduate years. The content we specify can be used to stimulate a search for the forms and means of instruction that will provide the best learning opportunities for teachers. It is for government officials who must arrange the proper conditions for good teacher preparation as well as professors and committees in schools and departments of education.

This book is for absolutely anyone who can make a positive impact, small or large, on the professional development of teachers during their in-service years. It is for school, district, and state employees who administer professional development programs as well as researchers, educators, and publishers' agents who provide professional development sessions.

Often the most influential decisionmakers are principals, school board members, and officials from local and state education agencies who provide the vision, time, and money necessary for in-service courses. Parents can, and should, participate in this process, such as the active PTA (parent-teacher association) members who raise private funds for teachers who want to take continuing education classes.

Finally, this book is for teachers themselves—from preschool through grade 4—who are committed to doing the best for their students and who are seeking the enriching programs and on-the-job experiences that will help them do it. It can be a prompt for study groups, a source of ideas to work through with facilitators, or a source of questions for coaches and mentors. In the increasingly competitive job market, when good teacher candidates are asked, "What inducements do you need to come to our school?," this book can help form the answer. At a minimum, teachers should request preparation and professional development to support them in their teaching of reading to every child in their charge.

we must do it, and we can[2]

We must act now to improve the education opportunities for teachers wherever they learn—in formal courses or through less formal professional development.

One reason is remedial. Many of the nation's colleges of education are simply not providing sufficient high-quality preparation, and they are the first to attest to their limitations. One researcher found that most teachers of the primary grades took an average of only 1.3 courses in the teaching of reading. A more recent survey indicates a mean of 2.2 courses. Teacher educators want more. The survey found a considerable discrepancy between the importance that faculty attaches to more reading courses and the ratings they give their current programs. Even with the slight increase, the total time spent on preparing to teach reading is entirely inadequate.

TEACHERS WANT TO BE BETTER PREPARED

- Only 28 percent of teachers reported feeling very well prepared to use student performance assessment techniques; 41 percent reported feeling very well prepared to implement new teaching methods; 36 percent reported feeling very well prepared to implement state or district curriculum and performance standards.

- While 54 percent of the teachers surveyed taught limited-English-proficient or culturally diverse students and 71 percent taught students with disabilities, relatively few teachers who taught these students (about 20 percent) felt very well prepared to meet their needs. The teachers' feelings about preparedness did not differ by teaching experience.

SOURCE: L. Lewis, B. Parsad, N. Carey, N. Bartfai, E. Farris, & B. Smerdon, 1999. *Teacher Quality: A Report on the Preparation and Qualifications of Public School Teachers.* Washington, DC: U.S. Department of Education, National Center for Education Statistics.

[2]For detailed studies and reports related to this section, see the Resources concerning the link between teaching resources and student achievement.

Considering the number of children at risk for reading failure in America, it is crucial that colleges and universities do a better job.

The picture is no better for in-service teacher education. On the job, teachers do not feel well prepared. We know how to define high quality in professional development. But the percentage of teachers who participate in such quality sessions is too low. For example, only half of teachers surveyed reported participation in professional development experiences to deepen their content knowledge. But that's good news compared to the dismal rates for other quality indicators: less than a quarter of the teachers spent enough time, or learned with the people they teach with, or had classroom follow-up via study groups, mentors, or internships. Worst case is not just a scenario: Only 5 percent reported that their professional development engaged them in active analysis of teaching and learning.

WE NEED TEACHERS

Over the next decade the nation's schools will need to hire 2.2 million teachers, over half of whom will be first-time teachers. Many schools already face shortages of qualified teachers, especially in high-poverty communities and in subjects such as math and science. We must do more to attract talented Americans of all ages into teaching.

SOURCE: U.S. Department of Education, Teacher Quality Initiative. Available online at www.ed.gov/inits/teachers/archive-recruit.html.

Specifically about reading, the first results from a new longitudinal survey show that about a third of our second- and third-grade teachers are participating in multiyear professional development in reading, and a quarter more had the chance for ongoing activity at least throughout the 1999 school year. But between a quarter and a third are still getting one-shot sessions passing as professional development for the teaching of reading. That is *not* quality. And this in the face of mounting evidence that *quality* professional development improves classroom practices.

Calls for urgent action come from the new challenges in the classroom. Teachers have to succeed with populations of students who come from cultures they do not understand and who speak languages they do not know. More knowledge can help teachers do a better job. Better professional development may attract more teacher recruits who are born into the communities that give them the culture and language knowledge that today's classrooms call for.

The current political climate of accountability provides further impetus. If teachers are going to help their students do better on high-stakes tests, the teachers need to understand achievement in important domains like reading and how to ensure student success without simply teaching to the test.

Finally, unless we act to provide high-quality professional development for teachers, we will lose them to other workplaces. Opportunities for continued learning help teachers stay engaged, function as real professionals, and avoid burnout. If we are going to retain the millions of new teachers who will be recruited over the next 10 years, we had better ensure they are feeling effective and supported.

Better teacher education and professional development are costly up front. We know that school districts spend only 1 to 3 percent of their resources on professional development, as compared to much higher expenditures for on-the-job training in most corporations and in schools in other countries. There's another false economy in personnel costs when schools hire so few teachers that they must create work schedules that allow most elementary school teachers only 8.3 minutes of preparation time for every hour they teach. Some of the best forms of professional development involve team planning and reflective evaluation or coached planning. How can that be done with such a time budget?

Enhancing teacher quality is an investment in human capital that yields real and lasting results. The effect is even greater on students who are at risk of low achievement than it is on other students. One recent study of more than 1,000 school districts concluded that every additional dollar spent on more highly qualified teachers netted greater improvements in student achievement *than did any other use of school resources.*

By strengthening our nation's instruction system, we make the right commitment. By strengthening our teachers, we give children the chance they need to succeed. There is no reason why reading achievement should be inadequate in the United States. The good news is that change is possible and within our grasp.

notes

Background for the book: The Carnegie Corporation of New York awarded a grant to Dorothy Strickland and Catherine Snow to develop a consensus document and work with professional development institutes. They convened the rest of the New Brunswick Group—M. Susan Burns, Peggy McNamara, and Susan Neuman—to develop the document. When Susan Neuman accepted an appointment as an assistant secretary of the U.S. Department of Education, she reduced her involvement in the group and felt that her participation was too minimal to be listed as an author. We thank her for her early contributions and enthusiasm. Within the New Brunswick Group, Peg Griffin took on major responsibility for prodding contributors and drafting prose.

The work was also supported by the American Education Research Association through a grant to Catherine Snow, M. Susan Burns, and Peg Griffin to work on the issue of teacher preparation in reading for classroom teachers of the early grades with Dorothy Strickland, Jean Clandinin, Linnea Ehri, Claude Goldenberg, and Annemarie Sullivan Palincsar. Laura Schenone worked with us on the development of this book, bringing the skills of a professional writer and the perspective of a concerned citizen and parent of young children. We appreciate her expertise and help. Jean Clandinin, Claude Goldenberg, and Annemarie Sullivan Palincsar reviewed a draft of the manuscript and gave us feedback for revision, as did Victoria Purcell Gates, Judith Green, Carol N. Dixon, and Sabrina Tuyay. We are grateful for their kind attention and absolve them of any responsibility for our failing to follow their advice on some points.

We also thank the teachers who informed us of their experiences and lent us their wisdom as well as their voices: Joan Gottesman, Bernadine Hansen, Debbie Johnson, Kia Martin, Yolene Medard, Ruth Nathan, Patrick Proctor, and Debra Weck.

The other members of the original Committee on the Prevention of Reading Difficulties in Young Children are Marilyn J. Adams, Barbara T. Bowman, Barbara Foorman, Dorothy Fowler, Claude N. Goldenberg, Edward J. Kame'enui, William Labov, Richard K. Olson, Annemarie Sullivan Palincsar, Charles A. Perfetti, Hollis S. Scarborough, Sally Shaywitz, Keith Stanovich, Sam Stringfield, and Elizabeth Sulzby. They continue to work toward improving children's reading achievement in different venues. Although they are not responsible for this book, their earlier work with us influenced it and we appreciate that and them.

resources

Basic reviews of the knowledge base relevant to learning to read in early childhood can be found in the following recent publications:

Kamil, M. L., Mosenthal, P. B., Pearson, P. D., & Barr, R. (Eds.). 2000. *Handbook of Reading Research: Volume III.* Mahwah, NJ: Erlbaum.

National Reading Panel. 2000. *Teaching Children to Read: An Evidence-Based Assessment of the Scientific Research Literature on Reading and Its Implications for Reading Instruction.* Washington, DC: National Institute of Child Health and Human Development.

National Research Council. 1998. *Preventing Reading Difficulties in Young Children.* Committee on the Prevention of Reading Difficulties in Young Children, C. E. Snow, M. S. Burns, and P. Griffin, Eds. Washington, DC: National Academy Press.

National Research Council. 1999. *Starting Out Right: A Guide to Promoting Children's Reading Success.* Committee on the Prevention of Reading Difficulties in Young Children, M. S. Burns, C. E. Snow, and P. Griffin, Eds. Washington, DC: National Academy Press.

Neuman, S. B., & Dickinson, D. K. (Eds.). 2001. *Handbook of Early Literacy Research.* New York: Guilford Press.

Partnership for Reading. 2001. *Put Reading First: The Research Building Blocks for Teaching Children to Read.* Washington, DC: National Institute for Literacy, National Institute of Child Health and Human Development, and U.S. Department of Education.

Standards for teachers are discussed from a variety of perspectives and change over time. Developing Educational Standards (*http://edstandards.org/StSu/Teaching.html*) is a web site maintained by Charles Hill and the Wappingers Central School District in New York. The site has links to standards proposed by professional organizations (e.g., *http://www.reading.org/advocacy/standards* for the International Reading Association) and those proposed by different states (e.g., for California, *http://www.ctc.ca.gov/profserv/progstan.html*). Organizations addressing novice and more advanced teachers both track and influence developments about standards related to the teaching of reading (National Board for Professional Teaching Standards, *http://www.nbpts.org;* Interstate New Teacher Assessment and Support Consortium, *http://www.ccsso.org/intaspub.html*). See also:

Darling-Hammond, L. 2000a. Transforming Urban Public Schools: The Role of Standards and Accountability. Paper presented at a conference entitled "Creating Change in Urban Public Education," December 7, The Joblessness and Urban Poverty Research Program, Harvard University. Available online at *www.ksg.Harvard.Edu/juprp/Sitepages/UrbanSeminar/UrbanEd/standards.pdf.*

The link between teaching resources and student achievement is investigated in such publications as the following:

Anderson, L., Evertson, C., & Brophy, J. 1979. An experimental study of effective teaching in first-grade reading groups. *Elementary School Journal, 79,* 193-223.

Darling-Hammond, L. 2000b. Teacher quality and student achievement: A review of state policy evidence. *Educational Policy Analysis Archives,* 8(1), 1-42.

Garet, M., Birman, B., Porter, A., Desimone, L., & Herman, R. (with K. Y. Soon). 1999. *Designing Effective Professional Development: Lessons from the Eisenhower Program.* Washington, DC: U.S. Department of Education, Planning and Evaluation Service.

Greenwald, R., Hedges, L. & Laine, R. 1996. The effect of school resources on student achievement. *Review of Educational Research, 66,* 361-396.

Porter, A., Garet, M., Desimone, L., Soon, K. Y., & Birman, B. 2000. *Does Professional Development Change Teaching Practice? Results from a Three-Year Study.* Washington, DC: U.S. Department of Education, Planning and Evaluation Service.

Roller, C. M. (Ed.) 2001. *Learning to Teach Reading: Setting the Research Agenda.* Newark, DE: International Reading Association. (See especially pp. 32-37).

Sanders, W. L., & Rivers, J. C. 1996. *Cumulative and Residual Effects of Teachers on Future Student Academic Achievement.* Knoxville: University of Tennessee, Value-Added Research and Assessment Center.

Wharton-MacDonald, R., Pressley, M., & Hampston, J. M. 1998. Literacy instruction in nine first-grade classrooms: Teacher characteristics and student achievement. *The Elementary School Journal, 99,* 101-128.

Teacher education and professional development are the topics of an ever-growing body of literature. While there are few absolutely compelling studies, some recent overviews can be consulted. The first is an attempt at a metanalysis specifically about preparation for teaching reading. The second and third are about studies of teacher learning but not specific to reading.

National Reading Panel. 2000. Teacher Education and Reading Instruction. In *Teaching Children to Read: An Evidence-Based Assessment of the Scientific Research Literature on Reading and Its Implications for Reading Instruction: Report of the Subgroups.* Washington, DC: National Institute of Child Health and Human Development.

Wilson, S., & Berne, J. 2000. Teacher learning and the acquisition of professional knowledge: An examination of research on contemporary professional development. *Review of Research in Education, 24,* 173-209.

Wilson, S. M., Floden, R. E., & Ferrini-Mundy, J. 2001. *Teacher Preparation Research: Current Knowledge, Gaps, and Recommendations.* Seattle, WA: Center for the Study of Teaching and Policy.

Organizations with different perspectives have addressed teacher preparation. The first two publications below specifically concern the teaching of reading:

Learning First Alliance. 2000. *Every Child Reading: A Professional Development Guide.* Baltimore: Association for Supervision and Curriculum Development.

Moats, L. 1999. *Teaching Reading IS Rocket Science: What Expert Teachers of Reading Should Know and Be Able to Do.* Washington, DC: American Federation of Teachers.

Other examples not specifically about reading include:

American Federation of Teachers. 2000. *Building a Profession: Strengthening Teacher Preparation and Induction.* Report of the K-12 Teacher Education Task Force. Washington, DC: Author.

Darling-Hammond, L. (Ed.). 2000c. *Studies of Excellence in Teacher Education.* Washington, DC: National Commission on Teaching and America's Future, American Association for Colleges of Teacher Education.

Fideler, E., & Haselkorn, D. 1999. *Learning the Roles: Urban Teacher Induction Practices in the United States.* Belmont, MA: Recruiting New Teachers.

Finn, C. E., Jr., Kanstoroom, M., & Petrilli, M. J. 1999. *The Quest for Better Teachers: Grading the States.* Washington, DC: Thomas B. Fordham Foundation.

Fullan, M., Galluzzo, G., Morris, P., & Watson, N. 1998. *The Rise and Stall of Teacher Education Reform.* Washington, DC: American Association of Colleges for Teacher Education.

Grossman, P., Thompson, C., & Valencia, S. 2001. *District Policy and Beginning Teachers: Where the Twain Shall Meet.* Seattle, WA: Center for the Study of Teaching and Policy.

Hirsch, E., Koppich, J. E., & Knapp, M. S. 2001. *Revisiting What States Are Doing to Improve the Quality of Teaching: An Update on Patterns and Trends.* Seattle, WA: Center for the Study of Teaching and Policy.

National Alliance of Business. 2001. *Investing in Teaching.* Washington, DC: Author.

National Commission on Teaching and America's Future. 1996. *What Matters Most: Teaching for America's Future.* New York: Author.

Wenglinsky, H. 2000. *How Teaching Matters. Bringing the Classroom Back Into Discussions of Teacher Quality.* Princeton, NJ: Educational Testing Service.

Handbooks and yearbooks include overviews from a variety of perspectives on teacher preparation:

Darling-Hammond, L., & Sykes, G. (Eds.). 1999. *Teaching as the Learning Profession: Handbook of Policy and Practice.* San Francisco: Jossey-Bass.

Griffin, G. (Ed.). 1999. *The Education of Teachers: Ninety-Eighth Yearbook of the National Society for the Study of Education.* Chicago: University of Chicago Press.

Iran-Nejad, A., & Pearson, P. D. (Eds.). 1999. *Review of Research in Education, Vol. 24.* Washington, DC: American Educational Research Association.

Kamil, M., Mosenthal, P., Pearson, P. D., & Barr, R. (Eds.). 2000. *Handbook of Reading Research, Vol. 3.* Mahwah, NJ: Erlbaum.

Murray, F. (Ed.). 1995. *The Teacher Educator's Handbook.* San Francisco: Jossey-Bass.

Richardson, V. (Ed.). 2001. *Handbook of Research on Teaching, 4th Edition.* Washington, DC: American Educational Research Association.

Sikula, T, Buttery, J., & Guyton, E. (Eds.). 1996. *Handbook of Research on Teacher Education, 2nd Edition.* New York: Macmillan.

Edited collections also address the issue of teacher preparation:

Cohen, D., McLaughlin, M., & Talbert, J. (Eds.). 1993. *Teaching for Understanding: Challenges for Policy and Practice.* San Francisco: Jossey-Bass.

Guskey, T. R., & Huberman, M. (Eds.). 1995. *Professional Development in Education: New Paradigms and Practices.* New York: Teachers College Press.

Lagemann, E. C., & Shulman, L. S. (Eds.). 1999. *Issues in Education Research: Problems and Possibilities.* San Francisco: Jossey-Bass and the National Academy of Education.

McLaughlin, M., and Oberman, I. (Eds.). 1996. *Teacher Learning: New Policies, New Practices.* New York: Teachers College Press.

Osborn, J., & Lehr, F. (Eds.). 1998. *Literacy for All: Issues in Teaching and Learning.* New York: Guilford.

Richardson, V. (Ed.). 1994. *Teacher Change and the Staff Development Process: A Case in Reading Instruction.* New York: Teachers College Press.

Richardson, V. (Ed.). 1997. *Constructivist Teacher Education: Building New Understandings.* Washington, DC: Falmer Press.

Roth, R. (Ed.). 1998. *The Role of the University in the Preparation of Teachers.* New York: Routledge Falmer Press.

Articles and policy briefs that address controversies, theories, evidence, and approaches about teacher preparation include the following:

Ballou, D., & Podgursky, M. 1998. The case against teacher certification. *The Public Interest, 132,* 17-29.

Ballou, D., & Podgursky, M. 2000. Reforming teacher preparation and licensing: What is the evidence? *Teachers College Record, 102,* 28-56.

Borko, H., & Putnam, R. (1996). Learning to teach. Pp. 673-708 in D. Berliner & R. Calfee (Eds.), *Handbook of Educational Psychology.* New York: MacMillan.

Clandinin, D. J., & Connelly, F. M. 1996. Teachers' professional knowledge landscapes: Teacher stories—stories of teachers—school stories—stories of schools. *Educational Researcher, 25*(3), 24-30.

Cochran-Smith, M. 1991. Learning to teach against the grain. *Harvard Educational Review, 61,* 279-310.

Cochran-Smith, M., & Fries, M. K. 2001. Sticks, stones, and ideology: The discourse of reform in teacher education. *Educational Researcher, 30,* 3-15.

Corcoran, T. B. 1995. *Helping Teachers Teach Well: Transforming Professional Development.* Philadelphia: University of Pennsylvania.

Darling-Hammond, L. 2000d. Reforming teacher preparation and licensing: Debating the evidence. *Teachers College Record, 102,* 5-27.

Feiman-Nemser, S. 1998. Teachers as teacher educators. *European Journal of Teacher Education, 21*(1), 63-78.

Ferguson, P., & Womack, S. T. 1993. The impact of subject matter and education coursework on teaching performance. *Journal of Teacher Education, 44,* 55-63.

Garet, M. S., Porter, A. C., Desimone, L., Birman, B. F., & Yoon, K. S. 2001. What makes professional development effective? Results from a national sample of teachers. *American Educational Research Journal, 38,* 915–945.

Grossman, P. L. 1991. Overcoming the apprenticeship of observation in teacher education coursework. *Teaching and Teacher Education, 7,* 345-357.

Hoffman, J., & Pearson, P. D. 2000. Reading teacher education in the next millennium: What your grandmother's teacher didn't know that your granddaughter's teacher should. *Reading Research Quarterly, 35*(1), 28-44.

Huberman, M. 1989. The professional life cycle of teachers. *Teachers College Record, 91*(1), 31-57.

Joyce, B., & Showers, B. 1996. Staff development as a comprehensive service organization. *Journal of Staff Development, 17*(1), 2-6.

Lampert, M. 1985. How do teachers manage to teach? Perspectives on problems in practice. *Harvard Education Review, 55*(2), 178-194.

Le Fevre, D., & Richardson, V. 2000. *Staff Development in Early Reading Intervention Programs: The Facilitator.* Ann Arbor: University of Michigan.

Little, J. W. 1993. Teacher's professional development in a climate of educational reform. *Educational Evaluation and Policy Analysis, 15*(2), 129-151.

Richardson, V. 2001. Alexis de Toqueville and the dilemmas of professional development. Paper prepared for the Center for the Improvement of Early Reading Achievement at the University of Michigan, Ann Arbor. Available online at *www.ciera.org.*

Showers, J., Joyce, B., & Bennett, B. 1987. Synthesis of research on staff development: A framework for future study and a state-of-art analysis. *Educational Leadership, 45*(3), 77-87.

Wildman, T., Niles, J., Magliaro, S., & McLaughlin, R. A. 1989. Teaching and learning to teach: The two roles of beginning teachers. *The Elementary School Journal, 89*(4): 457-479.

Zeichner, K. 1999. The new scholarship in teacher education. *Educational Researcher, 28*(9): 4-15.

1

From Shopping Lists to Poetry

Forms and Functions of Written Language

The journey toward literacy begins early and covers diverse terrain. Long before children arrive at kindergarten, they find written language in the world around them. In early encounters—often in the warmth of a caretaker's lap—a baby gets early impressions about how books operate: Pages turn, there are delightful pictures, Grandma's finger moves along those squiggly black marks. The baby may also begin to notice that the language of books—when read aloud—sounds different from the talk of regular life. There is a certain music to it, a certain cadence.

The growing child becomes aware of print not just in books but everywhere in the world. Print comes on cereal boxes and menus. On television there are letters of the day featured on *Sesame Street* or *Between the Lions*. There they are again on the sign for a fast food restaurant or in the church newsletter that came in the mail. There is written language in toy advertisements, bus schedules, and shopping lists. Printed language provides many kinds of valuable information: what time the party starts, how to put the toy together, which foods to buy at the store.

Surrounded by written language, a child, for quite some time, can be vague about what is going on—not always certain that reading and writing differ from talking or drawing, not sure what it takes to do it like the grown-ups. The literacy journey has a few bumps in the road, a few surprising events, and, finally, the meeting with teachers who are prepared to help. At last the child takes purposeful strides toward a clear goal—independent reading and writing to meet demands and create opportunities in everyday life.

A big step for a future independent reader and writer comes when young children begin to play "let's pretend." The importance of play for young children is

Learning to Teach Through Play

Debbie Johnson, Special Education Preschool Teacher, Minnieville Elementary School, Dale City, Virginia

Twenty years ago I first went to school to become a kindergarten teacher. Back then we thought play was important for learning social skills. Now we know there's much more to it. We know that the teacher might get in there and play with the kids, showing them how to role play and how to develop language skills. The teacher I worked under last year was really great at that. She would go into the housekeeping area and say, "Let's have a tea party. Who's going to help me? Who's going to dress up?" She'd get kids to talk and build their literacy skills. Watching her, I learned how to do it myself. I became more comfortable reaching kids in this way.

Sometimes you make things up as you go along depending on the child's interest. Sometimes you are flexible with that plan, what kinds of materials to use—like taking tissue boxes and using them as skates on the floor.

One workshop I went to stressed the importance of keeping a notepad and writing materials at each play center: the block center, the housekeeping center, the art center, the listening station, and the science center. This way the children can either draw a picture of something they did or integrate writing into their play—like writing a pretend letter if they're playing post office, or making a shopping list, or drawing an animal or a plant. We have a butterfly garden, and the kids sometimes draw pictures of a caterpillar turning into a butterfly. They can write, if they know how to write, or dictate to a teacher what they know about it. I now do this in my class, and the kids really enjoy it.

much overlooked. And yet play is crucial for a number of reasons. Playing at being a grown-up who reads and writes is a part of becoming a reader and writer. Besides that, any kind of pretend play accustoms the child to using symbols—to the idea in general that one thing "stands for" another. When little ones act out a pretend pirate scene, a rolled-up paper can stand for the telescope. Using letters as symbols is a central part of learning to read and write. For reading in a language like English, letters on a page "stand for" the sounds in spoken words—reading researchers call this the alphabetic principle.

Not enough teachers have a chance to learn about the early development of reading and writing. Preschool and kindergarten teachers, in particular, should be better prepared to nourish the growth of literacy that occurs before formal reading instruction. They should learn how to teach during different kinds of activities with children.

In too many preschools, literacy efforts focus exclusively on reading fiction to the large group once or twice a day and quizzing the group on the "letter of the day." While these are undoubtedly essential elements of learning to read, such efforts are not enough to prepare children for the challenges ahead. Preschool teachers must give children opportunities to use play toward their literacy development. It is anything but child's play for an expert teacher to teach through play.

From prekindergarten through the elementary grades, children must learn about the immense variety in written language—what it looks like and how it works. Teachers must help students become adept—as both readers and writers—with increasingly sophisticated and varied forms of written language. Increasing sophistication is experienced over time: The near tragedy of the Velveteen Rabbit; the travails of the "fourth-grade nothing"; the sadness but triumph of Sarah, plain and tall; the poignancy in the collection of O'Henry short stories—each book opens wider the door to the next.

But there are genres beyond the narrative. Teachers have to learn to teach beyond the stereotypes even for beginners. Expository prose should be used. A beautifully illustrated book with essays about frogs found in different parts of the world can augment knowledge gained from experience with a class aquarium. The same class might make a science notebook about the changes in the aquarium and reread parts as new entries are made each week. A current estimate is that, on average, less than 6 minutes a day is spent on information texts in first grades. This has to change.

playing with symbols

To become readers, children must understand that letters have the job of being symbols for sounds. But the first revolutionary accomplishment is engaging with any symbols at all. In play, children invent the symbols. They control them. They make the symbols work and change them as needed.

Well-prepared teachers understand how symbols work in play and how adults can teach children to sustain their symbolic play. The classic case is the broomstick that becomes a horse. The children know it is two things at once—the symbol is itself (a broomstick) and the thing it symbolizes (a horse). When it is a horse, certain rules apply. It can be ridden, it may be fed, it can even run away. The children agree to the meaning and use of the symbols in play.

While pretending, children even get to pretend about literacy. When children begin drawing and making other marks on paper, the results are quite idiosyncratic. The child knows it's a picture of a favorite toy or that it is the phrase "big wheel," even though no one else recognizes what the child has drawn or written. Even if the product requires an interpreter, the experience is nonetheless valuable. When

Hard at Play

"What do you want?" asks 4-year-old Debra.

It is playtime in Ms. Helen's preschool class, and the housekeeping area has been transformed into a restaurant. There are menus, posters with blue plate specials, order pads, a phone, a message pad, miscellaneous writing materials, a chef's hat, and dress-up clothes for the customers.

"What do you want?" repeats Debra, impatiently. She is the waitress—order pad and pen in hand.

Carefully scanning the menu left to right and top to bottom, Cecily says, "I'll have a coke and fries to go. And some Eggos."

Debra promptly writes some squiggly marks in a column—a list—on her pad. She yells the order to the kitchen.

Meanwhile, Cecily quietly puts a plastic pizza slice into her purse.

Debra shouts, "You take that pizza out of your purse or you'll never come to this restaurant again!"

Cecily leaves indignantly.

Debra, now alone, switches gears and begins to write on another pad. Ms. Helen visits the center but says nothing, as Debra fills the page with elaborate lines and circles, mumbling all the while: "Once upon a time, the lady . . ." and ending with ". . . the lady pays. The End."

A few moments later, Debra runs to the pretend telephone, which is evidently ringing. She answers it, "writes" a phone message, then overzealously hands both the phone and the message to Ms. Helen, saying "Here, it's for you."

Ms. Helen thanks her. The message contains round curvey marks, but the only letter Debra uses is D, the one she knows and likes from her first name.

Years ago, when she first began as a preschool teacher, Ms. Helen may have been tempted to interrupt the play scenario with some one-on-one teaching, perhaps encouraging Debra to put more real letters in her writing. But now she sees things differently. Thanks to reading some

writing *their* way or reading *their* way, children find that symbols give them the power to express and even make meaning. In this way they are learning what it means to become a reader and writer.

By the time preschool and kindergarten teachers are on the job, they should have an understanding of this aspect of child development and they should know how to engage children in play environments that form a path to literacy. Teachers should know how to provide materials and encouragement so that children learn about print in play. What does a pretend clinic need for a doctor to write a prescription while a patient reads the magazines in the waiting room? How can the teacher entice the pretend construction worker to make signs for the businesses and buildings she has made with blocks? Can a teacher help a superhero intercept a letter from a villain to a confederate?

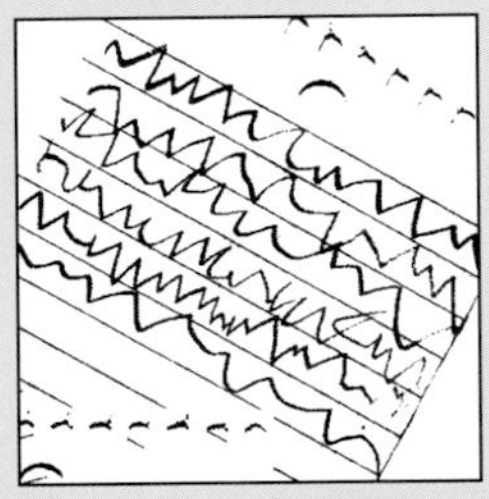

articles with the lead teacher who stepped in as her mentor, Ms. Helen understands that it is quite usual for 4-year-old children to "write" without conventional letters and that this type of play, in and of itself, shows progress toward becoming readers and writers. Debra's scribbles are longer and shorter depending on how much spoken language they are attached to—not yet the alphabetic principle, but a beginning connection between what's written and what's spoken.

Indeed, for this moment in their development, Ms. Helen thinks the girls are doing splendidly as they weave reading and writing activities into their play scenarios. At the end of the day, Ms. Helen looks over Debra's three writing products. Even with only the D as a recognizable letter, it's clear that Debra has an excellent understanding of different kinds of writing. Debra knows that she can use literacy for many purposes: to remember, to organize, and to make transactions. Debra clearly understands that print carries meaning and is useful in daily activities.

Debra may not use letters, but she uses literate conventions just the same. When taking an order at her restaurant, she knew to make a list, marking one little "word" under another. In contrast, when she worked on her story, she filled the page from edge to edge—left to right and top to bottom, showing knowledge of an important concept of print. And when taking a phone message, she attempted to use a cursive-like form. In each case, her writing changed to meet the task.

During their free-play time, Ms. Helen will continue to encourage her students to pretend to do the things grown-ups see as ordinary activities. Playing out the roles and rules expected in a restaurant, a home or a hospital—these are chances to develop background knowledge that later reading and writing can take advantage of.

Tomorrow, Ms. Helen will ask Debra to "read" back her page-long story. She is certain it will have a beginning, a conflict, and an end—as most good narratives do. What's more, there will even be a moral to this story: Pay up! Do not just take pizza slices in restaurants!

Play is also a way to work on understanding a story that has been read aloud to the children. Children might dedicate a particular scarf as the costume for a certain story, and every time they see that scarf they resurrect the story for days and days, increasing their familiarity with the language, realizing more about the meaning. All good early childhood teachers learn how to make sure children get the most out of a story by helping them act it out.

Teachers know, too, that playtime is a story waiting to be written. With the right materials and encouragement, a child becomes an author, writing as best as he or she can with crayons or markers or dictating to the teacher what happened to the pretend family in the housekeeping corner or to the truck drivers in the block corner. Sometimes the story is routine, sometimes exciting—but it's always good to take home or to pin up on the bulletin board to be read again later.

Teachers should learn to expect and encourage very young children to mix drawing, writing, playing, and speaking and to mix the symbols of pretend and conventional writing along with "estimated" spellings. (See Chapter 3, pp. 94-95, on the relationship between estimated spelling and phonemic awareness.) Eventually, it is the teacher's job to help children move toward conventional reading and writing, conforming by using only the letters of the alphabet and using them in the ways our spelling rules call for. An M may be the only attractive letter to a child named Margaret who loves Mom and likes mountains. But, after all, M cannot do the job that an O or an S can. The well-prepared teacher helps Margaret use "her" letter for writing but also helps her enlarge her letter knowledge.

Teachers must know how to arrange experiences so that children see the payoff for conformity to conventional use of the alphabetic symbol system. Through reading and writing we can communicate in many ways, over long periods of time and distances, with people we haven't ever seen or heard. It is almost magic that a child can read what his mother wrote when she was a little girl! What's more, a child can write a note so that her dad can find out what happened today even if he comes home so late everyone is asleep. And there are cousins far away who might respond to riddles written on a postcard.

Writing helps us remember, organize, conduct business, create, and appreciate imaginary worlds. And it all begins with symbols.

the workings of print

Teachers are often good at teaching basic concepts of print while reading to or with young children. They model and check on children's awareness of left-to-right eye or finger movements, sweeps to the beginning of a new line, and page turning when the last bottom right word has been read. They make sure the children know when they are reading (and relying on letters) and when they are talking about the pictures that enrich the text. They occasionally pause and subtly call children's attention to the spaces between words or point out punctuation marks or capitalization of letters. They readily take a moment to note the special size, alignment, and typefaces that set titles off from the rest of the page.

But what about print that doesn't involve our familiar 26 letters? Dozens of writing systems are used for business and government in the world today. Some conventions require people to read from right to left. Some readers must first move vertically down a page, rather than horizontally across it. Sometimes words are not separated by spaces. Some systems require symbols to change their shape depending on which symbols are next to them or whether they are at the beginning, middle, or end of the word. Our system does this a little bit in most styles of cursive

CONCEPTS OF PRINT FOR THE WORLDLY WISE

English is based on an alphabetic system. It relies on 26 letter symbols that represent more than three dozen phonemes, the smallest sound units in words. Spanish, French, and Italian are also alphabetic and rely on the same 26 signs (called the Latin alphabet), but writing these languages also calls for special symbols—called diacritics or accent marks. Russian readers and writers have 32 Cyrillic alphabetic symbols, some similar to the ones in the Greek alphabet. The Vietnamese writing system, Quoc Ngu, uses still other alphabetic signs.

In other parts of the world, people read and write with different basic units, not letters symbolizing phonemes. For example, in southern India, children learning to read Tamil use 246 symbols in a syllabic writing system. The Cherokee syllabary has 84 symbols. In syllabic systems, there is one written symbol for each syllable, not each phoneme.

Children from China are accustomed to a logographic system. In a logographic system there is one written symbol for each unit of meaning. In China, people need to recognize about 5,000 logographic characters to read a newspaper and about 30,000 to read literature. The same logographic characters can represent very different spoken words in the various languages used in China as well as Korea and Japan, which have adapted Chinese characters for their own use.

Some languages are written with systems that have more than one type of symbol. Japanese, for instance, uses both syllabic and logographic symbols.

writing. Some societies, like Norway, do not have conventionalized spelling—it's up to each writer to decide how to spell a word!

With increasingly multicultural and multilingual cachement areas for their schools, teachers should be prepared to work effectively with children who have quite different concepts of print than they do. Teachers need to know that the very idea of an alphabet may be entirely new to children who first encounter English as a second language but know about reading and writing in another language. Teachers must understand when literacy at home or church introduces expectations that written English does not meet. They must find ways to help children transfer what they can for use in reading and writing English and to add the new concepts of print they will need for English. Most important, they must know that their way isn't the only way.

Teachers need to teach more sophisticated print concepts as children get older and encounter more challenging texts in the second, third, and fourth grades. It is difficult for many children at this age to read nonfiction if they do not understand the conventions involved in chapter headings, subheadings, graphs, tables, illustrations, captions, indexes, or glossaries. Teachers need to be skilled at introducing these elements. Most important, teachers need be disposed to put these features into action in the daily life of the classroom.

Consider the concept of an index in a book, for example. Instead of saying, "Read pages 20 through 23 in the geography book," the knowledgeable teacher says, "Read about the spiders we might find in the United States." She then leads the class though problem solving that will show the difference (and different usefulness) between a table of contents and an index. They figure out whether to start with "spiders" or "United States." They persist when "United States" is not there with "spiders" and realize that "North America" will do. Finally, "Spiders, North America, 20-23." The result may be the same pages to read, but the bonus is an opportunity to practice an advanced print concept.

Teachers need to learn to introduce children to reference books and to remember to put them to work in the daily life of the classroom as well!

how texts vary

In addition to familiarity with different parts of a book or other written material, children must become familiar with different types of written material. In fact, children begin early to learn the conventions of different genres. "Once upon a time," for example, is well known early on as the signal for a fairy tale.

But less appreciated subtle understandings of genre differences also begin to develop early. Take menus and shopping lists, for example. When we see the words *pie* and *cake* written under each other on a shopping list, we understand that we are to get both. When we see the same words under one another on a menu, though, we know we should pick just one. The word "and" is not written on a shopping list; "and" is implied in the shopping list genre. The menu genre implies "or." These two genres are seldom taught explicitly, but they are understood by young children.

Teachers need to know about the early accomplishments and build on them to expand the children's repertoire—both about the signs to look for (like "once upon a time") and the understanding of what is meant even though it is not written in so many words (like "and" for shopping lists and "or" for menus).

Children may not expect the differences that genre brings in other cases, though. The content and objective factual tone of a news report about a county fair may be surprising to a child who connects that topic with lively stories of personal experiences. Descriptive passages from an omniscient viewpoint may similarly be unfamiliar to a child.

As children get into second grade and beyond, they spend more of their time reading to learn, rather than learning to read. For much of their educational lives, they will depend on informational expository texts—in science, social studies, all the different domains of learning. And yet, despite this reality, most of the reading time in the lower grades is spent on fiction and nonfiction narratives.

It is imperative that teachers have the opportunity to be prepared to do a better job teaching young children to read expository prose. Teachers should introduce and provide practice with useful procedures and techniques. Some are procedures that can be applied to many texts—teaching students to survey a whole chapter before reading to notice the main ideas organized by the subheadings, pictures, figures, or tables. Some techniques are more unusual—instructing students to make a glossary of their own by picking the terms in italics or bold print that indicate definitions.

Teachers need to apply what they know about reading different academic subjects. When we are able to read to learn we are fluent with the specialized written forms found in different subjects. Even in the lower elementary grades, textbooks and trade books have certain markers associated with, for example, history or science, that children can learn to rely on.

Children who have opportunities to work with print in a wide variety of forms and functions have the motives and means to succeed as readers. We know practice helps children become good readers. We also know that sometimes it is hard to get children motivated or to keep their attention during extended practice. But think about this immense variety of different forms of written language—here is an opportunity for multiple chances to practice. Here a news report, there a book of riddles, an essay, an invitation, a mystery, a poem—always new and different functions and forms and topics. But each passage provides practice with many of the same clauses, the same words, the same syllables, the same letter-sound correspondences. The variety means that practicing to full fluency need not be drill or dull.

oral and written language

Spoken and written language can be very different. Compare two reports of the same event:

1. He just wannenna chill 'n jam and the dweeb ups 'n disses 'm!

2. Although the sole intent of the young man appears to have been to relax at a party, he was criticized to the point of being denied due respect. The criticism was issued, without warning or notice, by the person (introduced above) who is widely believed to be unaware of the social norms upon which we depend for probity.

While these are clearly caricatures, they call attention to the wide range of contrasts between spoken and written language. Both utterances are contemporary American English. The first is clearly an example of spoken language and gives a hint about the speaker's youth. It is unlikely to be printed, except as a direct quotation. The vocabulary is colloquial: *chill, jam, up and, diss, dweeb.* Most teens can supply a translation. The past tense of the verb in the first clause slips into the "historical present" in the second. There are contracted pronunciations for prepositions, pronouns, and conjunctions—*wannenna* instead of wanted to, *'n* instead of *and, 'm* instead of *him.*

The second example has vocabulary and phrases more common in written language, such as *sole intent, due respect,* and *probity.* We also find hallmarks of formal syntax: passive constructions (*appears to have been* and *are widely believed to be*), explicitly limited claims, and subordinate clauses (introduced by *although, who,* and *which*).

An astute reader might recognize that *the* before *dweeb* in the first example translates into the *introduced above* in the written version. Had *a dweeb* been used instead, we would take it for granted that the character had not previously been mentioned in the narrative.

With these and many other sorts of contrasts, it is no wonder that the transition from oral language to written language can be difficult for children. Teachers must prepare to help children become familiar with the subtleties, tone, vocabulary, styles, and other conventions of written language. In the talk of everyday life, so much can be easily conveyed with tone of voice, gestures, pregnant pauses, posture, and sheer physical closeness. We can even stop speakers and get them to help us understand.

Written language on a page must rely on other tactics instead. Precise word choice, unusual grammatical structures, lengthy descriptive passages may just sound "fancy" or formal to the new reader, but teachers have to know how to help students understand the ways that these features express meanings when face-to-face oral-language tactics are not available.

Teachers need to be prepared to uncover the subtleties of written language on the spot and prepare children to solve such problems themselves as they read independently.

writing—when children control the symbols

Readers have one viewpoint on literacy and written language. Writers see it from the other side. The concepts about written language that we rely on for reading are the very same ones we must execute as writers. Neither viewpoint on literacy is complete. Each helps develop the other. And doing one often leads to doing the other.

As astute observers of their students' writing products, knowledgeable teachers find out a lot about where the students are as readers. But they must be well prepared to analyze this information. In preservice courses, teachers should obtain a good understanding about writing development in children and the appropriate demands to place on writing production for children of different ages.

Teachers also need to learn to develop good instructional plans for writing that support reading achievement. There are four features to take into account.

1. Writing as a Part of Daily Classroom Life

Teachers need to learn how to arrange writing practice during the school day, when they can prompt and motivate children, encourage them to stay on task, and give them the help they need to succeed. Effective teachers know that writing is not just for homework and that writing time is not necessarily quiet. Good writing can be part of a dynamic shared experience. A class writes a get-well letter to a sick peer. A pair of students brainstorm ideas for a science project. A team of writers collaborates to write, revise, and perform a play. Teachers also need to be prepared to manage a classroom so that space and time are available for the child whose writing blossoms in solitude.

2. Writing for a Reason and Purpose

Effective teachers know how to communicate to children that writing matters. A descriptive essay has a purpose if it describes the city zoo on a web site. A good short story deserves reading, and the author deserves the chance to promote it during an interview on a real or pretend TV talk show. Children notice that a lot of things are unfair from their point of view; who better to write a persuasive letter to a storekeeper or a mayor? And who better to follow that up with a letter of protest if there is no response? Effective teachers have learned that if letters are assigned, they must be stamped and sent, and correspondents should be chosen according to the likelihood that they will return mail in a timely fashion. Poetry slams have captivated middle school and high school students, and fourth graders might find them a good venue for their efforts. E-mail and distance collaboration projects for science or geography are important for teachers to learn about.

Of course, teachers should provide high-quality feedback about the writing, noticing improvements or the need for them. But giving a child's writing a grade or even extended commentary is not a substitute for arranging opportunities for purposeful writing.

3. Writing and the Teacher's Influence

Teachers should know how to gear writing assignments to showcase a child's accomplishments but should also pave the way for growth and improvement. Some moments are the right ones for a teacher to model good writing or to be a collaborator. Other times the teacher needs to stand back and let children write on their own. All along the way, he or she must make moment-to-moment judgments about what will be most helpful: to step in and closely direct children, to give them writing prompts or take dictation, to act as a word source, to point to a dictionary, to arrange peer writing groups, or to provide a sample for children to work from when they are experimenting with a new genre. That there is one tried-and-true approach is a myth.

Teachers need to be well prepared with techniques that can help children navigate the whole writing process. They should be skilled in helping children prepare to write, organize their thoughts and ideas, and discover when and how to revise.

4. Writing in Different Genres

The earliest composing that a child does is often a report of an event that just happened, dictated to a teacher. A good teacher must encourage children to move from personal narrative toward third-person narrative, to fiction and poetry of different styles, and to expository prose that informs or persuades, gives instructions or directions.

There is a lot for teachers to master about the forms and functions of symbols and written language. The payoff for the effort is when the teacher chooses curricula, plans lessons, responds to the unexpected question of a child, helps a child around a stumbling block or through a hard task. To do the simple daily acts, an effective teacher has an ever-growing knowledge base and the disposition to act on it and with it.

Activity

Child's Play

To take professional development on play seriously and make it effective, a teacher study group is helpful. This activity is designed to move between discussion of publications about literacy and play, through collaborative planning, to implementation and reflection on the effectiveness of the attempt. A facilitator familiar with the literature and the classroom is a helpful resource for a teacher study group. A sample unit plan for such professional development follows:

- Begin to discuss some background reading on the play/literacy connection with a group of teachers (from the same school or very close neighboring ones).

- Demonstrate the way reading and writing opportunities can enrich children's pretend play if they are woven into the action. Use a class visit with a teacher interview or a videotaped or written case example. (For example, a classroom might feature a veterinary hospital setting for pretend play. An enterprising teacher would take students on a field trip. Books about animals and vets would be featured in the library corner and be read at story time. Visitors who deal with sick animals would be invited to show and tell what they do. The teacher would gather and set up props for animal patients, owners, vets, and vet techs. The play center would include sufficient writing materials to create file records, phone messages, and prescriptions. Don't forget the magazines in the waiting room.) Refer back to the reading to discuss the demonstration.

- Model the ways that a teacher can help children become more adept at the roles and rules involved in this pretense. Point out strategies that make the play last, maximizing cooperation among the children and minimizing inattention and conflict.

- Have the teachers work in a few small groups to develop plans for other play opportunities that involve literacy. These plans should include not only the classroom setting, visit schedules, props, and writing and reading materials but also guidelines for the teacher's interactions with the children.

- Have the whole group evaluate the plans for adequacy and practicality and then revise as needed. Use background reading on the play/literacy connection to guide the evaluation and revision.

- Choose one of the plans to "pilot test" in at least one classroom. Give the teachers the opportunity to see it in action and assess whether or not the children are playing with symbols and developing their literacy skills.

- Repair and share—adjust the pilot situation to emphasize more engaging or lengthier play or more literacy, as needed. Find ways to share preparation time and materials so the different play scenarios can be cycled among classrooms.

Activity

Written English Is Alphabetic, BUT. . .

To evaluate curricula and materials and to interpret children's responses and overall progress, teachers rely on technical knowledge about written English. A teacher educator can make the dry technical terms more memorable and stimulate more discussion by weaving in some problem solving.

This activity is a guide to teaching teachers about the fact that symbols in a writing system have two parts: the written signs and the units in the language that are to be represented. The kind of unit represented identifies the type of writing system: *syllabic, logographic,* or *alphabetic.* It begins with general thought-provoking demonstrations and leads to problem solving and practice.

1. English is *alphabetic,* representing phonemes with letters. But there are no relations to phonemes in a written "26" or "80" (even though the phonemes are represented by letters if we write out the words "twenty-six" and "eighty").

   - Have the class find people who will read a list of numbers in Spanish, French, German, Russian, and any other languages in the neighborhood. Report the results to the class. Note that there are all sorts of different sounds, different phonemes!

   - Get a French speaker to interview in the class. Have the speaker read "80" for the class and then provide the literal English translation of what is said. It isn't like "eight"; it's like Lincoln's Gettysburg address—"four score!"

   - Discuss the limited way in which English is a mixed system—mostly *alphabetic,* but we can think of our use of numbers as *logographic.*

2. History, geography, politics, accident, and ease—all play a role in determining which writing system is actually used for a language. (In fact, the Serbs and Croats share a language, but each uses a different alphabet to write it.) Suppose English was written with a syllabary or a logographic system. How many symbols would we use to write some simple sentences?

Write the two following simple sentences on the board:

I like radio. **I hate ice cream.**

Ask the class to discuss how many symbols it would take to write each sentence if we wanted to use a syllabary. In a *syllabic* system, there would be one sign for each syllable.

Five signs for "I-like-ra-di-o" **Four signs for "I-hate-ice-cream"**

Ask the class to discuss how many symbols it would take to write each sentence if we wanted to use a logographic system. In a *logographic* writing system, there would be one sign for each unit of meaning (technically called a morpheme).

Three signs for "I + like + radio" **Three signs also for " I + hate + ice cream"**

Expect some controversy here. Two words aside, "ice cream" is only one morpheme. Something like "heavy cream" has two morphemes; one refers to the basic substance—cream—and another limits the reference to a particularly dense type of cream. But "ice cream" is different; it is not just a kind of cream that is icy. (Thank heavens!)

Ask the class to discuss how many symbols it would take to write each sentence if we wanted to use an alphabetic system. In an alphabetic writing system, there would be one sign for each phoneme (the smallest unit of sound). It is helpful to provide the students with a chart or table listing the phonemes of English with key words as a pronunciation guide. Be very clear that the task is to count the phonemes *not* the letters.

Nine signs for /ay/ /l/ /ay/ /k/ /r/ /ey/ /d/ /iy/ /ow/ (I like radio)

Ten signs for /ay/ /h/ /ey/ /t/ /ay/ /s/ /k/ /r/ /iy/ /m/ (I hate ice cream)

Expect an uproar. Some students may still have trouble reporting how many phonemes there are in a word. It's a good deal harder than for syllables. For others that's not a problem, but they want the chance to complain about conventional spelling.

There are only 26 letters in the alphabet that we use to write English, but spoken English has more than three dozen phonemes. Some doubling up with special orthography (spelling rules) is called for. English readers know that "silent e" influences the pronunciation of the first vowels in *like* and *hate* and that the digraph *ea* represents the single long vowel in the spoken word *cream.*

An alphabetic system that has these complications is said to have a *deep orthography.* Some languages, like Spanish, have conventional spellings more directly connected to the surface sounds of the contemporary spoken language.

Ask each student to make one sample sentence for practice on counting units. Pair them up to repeat what you just did with the whole class. Have half put their sentences in a bowl; have the others pick one and find the author, so they can work together on both samples.

Reconvene the class to discuss samples the students disagree about or think are interesting. Take the opportunity to note the growth of knowledge of the technical terms in action in teacher-to-teacher talk.

Activity

Getting to Know a Reference Book

Using a reference book and being prepared to teach students to use it are quite different. A class or teacher study group can focus on one type of reference book and develop local expertise about it that can be available for teachers to design and implement lessons. The group will be better quipped to prepare lessons about other reference books, whether as a group or as individuals.

Divide the group into pairs or triplets and give each a thesaurus. Ask them to study the book and observe its features. What kinds of words are in boldface? Which words are in italics? How are numbers used? Are entries organized alphabetically or by subject? Is there an index?

Ask teachers to write their answers and then discuss them with the larger group. When agreement about the key features has been reached, brainstorm about introducing students to reference books and providing practice so they can use the key features effectively. Try out the ideas in role play, using appropriate age-level thesauruses, dictionaries, or encyclopedias.

Activity

Science Does . . . History Does . . .

Many teachers have had little chance to study how texts in different content areas are in fact different. How, then, can they help children apply and extend their reading skills across the curriculum?

A teacher study group can spend a year "attacking" academic subjects two at a time. The teachers can identify the special tricks for reading well in each subject area. They can determine when and how this information can become lessons with children. The following activity would span about 6 weeks:

- Divide into pairs. Each pair takes three samples from a science book intended for third or fourth graders and three samples from a history book intended for the same audience. To be manageable and productive the samples should be about four or five pages long.

- Spend at least two weekly hour-long meetings studying the passages to find the elements that recur in the samples. Here are some issues to address:

  List features of layout and language that a reader can expect: Do the subheads seem to all work in the same way? Are there sidebars? Illustrations? Graphics and font differences? Tables and charts? Bulleted lists? Are definitions indicated? If so, how? Are examples indicated? If so, how?

  Make a generalized outline that just about always predicts the kind of information that will be presented at certain points in a chapter.

  List a set of questions that just about always get answered.

  Is there an argument structure that comes up again and again (similarities and differences, cause and effect, etc.)?

  How is evidence presented? Where do conclusions come in? What happens to uncertainties?

  What are the standard opening and ordinary closing for new topics?

  What's taken for granted and thus seldom said? What sort of information is usually stated outright and what sort usually has to be figured out or looked up somewhere else?

- After the 2 weeks of collecting information on the samples, bring together the larger group for two meetings and pool the findings. Then sort them: What does science always do? Or almost always? Does history never do it? Or hardly ever? And vice versa. What features are shared by the disciplines? List the findings about science. List the findings about history.

- In the next meeting, consider the lists in light of the children and the curriculum demands. Should the information be used with the children? If so, specifically when and how? Should it be incidental teaching as opportunities or problems arise? Should it be intentional lessons? How can the teachers evaluate the plans and their implementation to see if they promote fluent reading and lasting comprehension of science and history texts? The group may decide to collaborate on lesson plans, pilot implementations, and evaluations and repairs of them. The group may decide to establish an archive of the information and plans to share with other teachers.

- Eventually, a similar study of two different curriculum areas can be undertaken.

Activity

How Written Language Works

Teachers may be unprepared to bridge the gap between the language of conversations and written language. This activity gives them a chance to appreciate the features of written language that might be unfamiliar enough to become a stumbling block for some of their students.

Have teachers work in small groups. Give each group some children's books of varying types—classics of children's literature; easy readers; others that feature popular television, comic, or music characters.

Ask the teachers to identify the vocabulary, idioms, grammatical constructions, and stylistic devices that are unusual in spoken language but common in books.

Discuss (1) how to prepare children to understand a passage despite any unusual features and (2) how to engage children in thought about the unusual features of written language, so that they can handle similar examples when reading on their own.

A B C D E F G H I J K L M N O P Q R S T U V W X Y Z

Activity

Writing in Many Genres

Teachers, like other literate adults, can take for granted the knowledge they have about different genres of written language. This activity is meant to help them bring this knowledge to the surface and have it become a resource for their students.

1. Ask teachers to consider the ordinary experience of night turning into day. Then have them write accounts or explanations of this phenomenon in four different genres: a folkloric or oral tradition, a scientific explanation, a narrative description, and a poem describing any aspect of the phenomenon.

   This activity should be lighthearted and fun. Have teachers pass their creations around the class or read them aloud to the class.

   - Ask teachers to brainstorm a list of features that characterize the genres they just worked with.
   - Ask teachers to discuss ways to adapt this activity for use in their classrooms.

2. Ask teachers to think about what genres are by transforming them!

   - Pair up the members of the class. Pass out copies of a local news story and copies of the lyrics from a currently popular song.
   - Give them a half hour to make a song from the news report and a news report from the song.
   - Pass around the "new genres." Ask for volunteers to read their efforts aloud. Discuss what the teachers had to know about news reports and songs to do the task.
   - Next ask the class to think about a favorite poem and a favorite novel. Discuss which aspects of a novel could be captured in a poem and which couldn't. Are there any novels or poems that evoke similar responses in readers? Are there any themes or topics that would be especially unlikely to turn up in a novel or a poem?

3. Ask the class to think about both exercises. Do they think differently now about genres? What have they learned as readers that their students as beginners may not yet know? How can their knowledge be woven into their planned activities with students?

resources

Basic reviews of the knowledge base relevant to this chapter can be found in the following recent publications:

Kamil, M. L., Mosenthal, P. B., Pearson, P. D., & Barr, R. (Eds.). 2000. *Handbook of Reading Research: Volume III.* Mahwah, NJ: Erlbaum. (See especially the chapter by Yaden, Rowe, and MacGillivray on emergent literacy.)

National Reading Panel. 2000. *Teaching Children to Read: An Evidence-Based Assessment of the Scientific Research Literature on Reading and Its Implications for Reading Instruction: Report of the Subgroups.* Washington, DC: National Institute of Child Health and Human Development. (See especially Chapter 4, Part II, "Text Comprehension Instruction.")

National Research Council. 1998. *Preventing Reading Difficulties in Young Children.* Committee on the Prevention of Reading Difficulties in Young Children, C. E. Snow, M. S. Burns, and P. Griffin, Eds. Washington, DC: National Academy Press. (See especially Part I, Chapter 2, and Part III.)

Neuman, S. B., & Dickinson, D. K. (Eds.). 2001. *Handbook of Early Literacy Research.* New York: Guilford Press. (See especially the chapter by Whitehurst and Lonigan on emergent literacy and the one by Dyson on writing and symbols.)

The sections of this chapter can be elaborated by consulting various other sources, including the following:

Playing with symbols:

Bloodgood, J. W. 1999. What's in a name? Children's name writing and literacy acquisition. *Reading Research Quarterly, 34*(3), 342-367.

Bodrova, E., Leong, D. J., Hensen, R., & Henninger, M. 2000. Imaginative, child-directed play: Leading the way in development and learning. *Dimensions of Early Childhood, 28*(4), 25-30.

Clay, M. 1975. *What Did I Write?* Auckland, New Zealand: Heinemann.

Dever, M. T., & Wishon, P. M. 1995. Play as a context for literacy learning: A qualitative analysis. *Early Child Development and Care, 113,* 31-43.

Dickinson, D. 1996. *Emergent Literacy and Dramatic Play in Early Education.* Albany, NY: Delmar.

Ferreiro, E., & Teberosky, A. 1982. *Literacy Before Schooling.* Exeter, NH: Heinemann Educational Books.

Gundlach, R. 1982. Children as writers: The beginnings of learning to write. Pp.129-148 in M. Nystrand (Ed.), *What Writers Know.* New York: Academic Press.

Neuman, S. B., & Roskos, K. 1992. Literacy objects as cultural tools: Effects on children's literacy behaviors in play. *Reading Research Quarterly, 27,* 202-225.

Pellegrini, A. D., & Galda, L. 1993. Ten years after: A reexamination of symbolic play and literacy research. *Reading Research Quarterly, 28*(2), 162-175.

Purcell-Gates, V. 1996. Stories, coupons, and the "TV Guide": Relationships between home literacy experiences and emergent literacy knowledge. *Reading Research Quarterly, 31*(4), 406-428.

Strickland, D. S. 1991. Emerging literacy: How young children learn to read. Pp. 337-344 in B. Persky & L. H. Golubchick (Eds.), *Early Childhood Education, 2nd ed.* Lanham, MD: University Press of America.

Sulzby, E. 1985. Children's emergent reading of favorite storybooks: A developmental study. *Reading Research Quarterly, 20*(4), 458-481.

Teale, W. H. 1987. Emergent literacy: Reading and writing development in early childhood. Pp. 45-75 in E. Readance and R. S. Baldwin (Eds.), *Research in Literacy: Merging Perspectives.* Thirty-sixth yearbook of the National Reading Conference. Rochester, NY: National Reading Council.

Teale, W. H., & Sulzby, E. (Eds.). 1986. *Emergent Literacy: Writing and Reading.* Norwood, NJ: Ablex.

Teale, W. H., & Sulzby, E. 1989. Emergent literacy: New perspectives. In D. S. Strickland and L. M. Morrow (Eds.), *Emerging Literacy: Young Children Learn to Read and Write.* Newark, DE: International Reading Association.

The workings of print

Coulmas, F. 1996. *The Blackwell Encyclopedia of Writing Systems.* Oxford: Blackwell.

Daniels, P. T., & Bright, W. (Eds.). 1996. *The World's Writing Systems.* New York: Oxford University Press.

Sampson, G. 1985. *Writing Systems.* Stanford, CA: Stanford University Press.

How texts vary

Armbruster, B. B., & Anderson, T. H. 1984. Structures of explanations in history textbooks or so what if Governor Stanford missed the spike and hit the rail? *Journal of Curriculum Studies, 16*(2), 181-194.

Caswell, L., & Duke, N. 1998. Non-narrative as a catalyst for literacy development. *Language Arts, 75,* 108-117.

Doiron, R. 1994. Using nonfiction in a read-aloud program: Letting the facts speak for themselves. *The Reading Teacher, 47,* 616-624.

Duke, N. 2000. 5.6 minutes per day: The scarcity of informational texts in first grade. *Reading Research Quarterly, 35*(2), 202-225.

Hepler, S. 1998. Nonfiction books for children: New directions, new challenges. Pp. 3-17 in R. A. Bamford & J. V. Kristo (Eds.), *Making Facts Come Alive: Choosing Quality Nonfiction Literature K-8.* Norwood, MA: Christopher-Gordon.

Kamberelis, G. 1999. Genre development and learning: Children writing stories, science reports, and poems. *Research in the Teaching of English, 33*(4), 403-460.

Leal, D. 1993. Storybooks, information books, and informational storybooks: An explication of the ambiguous grey genre. *The New Advocate, 6,* 61-70.

Meyer, B. J. 1975. Identification of the structure of prose and its implications for the study of reading and memory. *Journal of Reading Behavior, 7*(1), 7-47.

Pappas, C., & Barry, A. 1997. Scaffolding urban students' initiations: Transactions in reading information books in the read aloud curriculum. Pp. 215-236 in N. J. Karolides (Ed.), *Reader Response in Elementary Classrooms: Quest and Discovery.* Mahwah, NJ: Erlbaum.

Strickland, D. S., Ganske, K., & Monroe, J. K. 2002. Improving reading comprehension. Pp. 141-154 in *Supporting Struggling Readers and Writers: Strategies for Classroom Intervention 3-6.* Portland, ME: Stenhouse.

Oral and written language:

Biber, D., & Finegan, E. (Eds.). 1995. *Sociolinguistic Perspectives on Register.* New York: Oxford University Press.

Duke, N. K., & Kays, J. 1998. "Can I say 'once upon a time'?": Kindergarten children developing knowledge of information book language. *Early Childhood Research Quarterly, 13*(2), 295-318.

Kavanagh, J. F., & Mattingly. I. G. (Eds.). 1972. *Language by Ear and by Eye: The Relationships Between Speech and Writing.* Cambridge, MA: MIT Press.

Ochs, E. 1979. Planned and unplanned discourse. Pp. 51-80 in *Discourse and Syntax.* New York: Academic Press.

Roberts, B. 1992. The evolution of the young child's concept of "word" as a unit of spoken and written language. *Reading Research Quarterly, 27*(2), 124-138.

Tannen, D. (Ed.). 1982. *Analyzing Discourse: Text and Talk.* Washington, DC: Georgetown University Press.

Writing: Because the emphasis here is on reading, our coverage of writing development and teaching is limited. The following publications are a few starting points for more adequate coverage:

Applebee, A. N. 1986. Problems in process approaches: Toward a reconceptualization of process instruction. Pp. 95-113 in A. R. Petrosky and D. Bartholomae (Eds.), *The Teaching of Writing: Eighty-Fifth Yearbook of the National Society for the Study of Education, Part II.* Chicago: University of Chicago Press.

Dyson, A. H. 1988. Unintentional helping in the primary grades: Writing in the children's world. Pp. 218-248 in B. A. Rafoth & D. L. Rubin (Eds.), *The Social Construction of Written Communication.* Norwood, NJ: Ablex.

Graves, R. L. (Ed.). 1999. *Writing, Teaching, Learning: A Sourcebook.* Portsmouth, NH: Boynton/Cook.

Jensen, J. M. 1993. What do we know about the writing of elementary school children? *Language Arts, 70,* 290-294.

Pressley, M., Wharton-McDonald, R., Allington, R., Block, C. C., Morrow, L., Tracey, D., Baker, K., Brooks, G., Cronin, J., Nelson, E., & Woo, D. 2001. A study of effective first-grade literacy instruction. *Scientific Studies of Reading, 5*(1), 35-58.

Strickland, D. S., Ganske, K., & Monroe, J. K. 2002. Improving writing. Pp. 167-199 in *Supporting Struggling Readers and Writers: Strategies for Classroom Intervention 3-6.* Portland, ME: Stenhouse.

2

Making Meaning

Language Development and Comprehension

We read for meaning. Whether the text before us offers poetry, scientific facts, a page-turning plot, or the most mundane instructions about how to put together a set of shelves, a basic goal is that we, the readers, understand what the writer intended. We may also enjoy it, disagree with it, sneak to the last page, suddenly have a revolutionary new idea beyond anything the author conceives, or respond in a number of other ways.

Good readers call on many resources to do all this. By the time we are adults, most of our comprehension work becomes so automatic that we are no longer aware of it. When we pick up a newspaper, we effortlessly understand words and phrases we've seen in myriad styles of print and writing and heard in countless conversations.

Even when faced with new subject matter, we draw resources from our old knowledge, our background. Of course, there are the well-developed connections between letters and sounds that carry our word knowledge from speaking and listening into reading and writing. But that's not all we exploit for comprehension. We capitalize on experience with all aspects of language—the routine structure of clauses and sentences, paragraphs, articles, and whole books. We draw expectations, analogies, and ideas from other subjects. We know enough to check on ourselves as we read along and even reread a page or two if we get confused, keeping careful track of how previous parts inform the later ones, alert for apparent contradictions that may mean we misunderstand or that the author is wrong!

For most beginning readers, it is a different story. Their resources are not as abundant; the work of comprehending is not as familiar. Understanding the written language is not so certain.

Most teachers are good enough at the age-old method of checking students' comprehension by asking questions at the end of a story. But it is *not* good enough for children if teachers only test reading comprehension rather than teach it.

Comprehension instruction proceeds on three fronts at once: resources, tactics, and repairs. Students build up their resources—the growth and elaboration of spoken language, the depth and breadth of concepts and vocabulary. They deploy their resources, becoming adept tacticians before and during reading. And students learn strategies to fall back on when meaning breaks down.

Teacher education colleges must do a better job of preparing aspiring teachers

Midstov

Mrs. Keane and her first-grade class are sitting on the rug while she reads to them after lunch, just as she does every day. She has an old book of tales that one of the families sent in. At other times in the day, what's read is more contemporary, and most of the children are more likely to read on their own than listen. But just before rest time they like this sort of book.

As she closes the book, she nods to Tyrone, who has been waiting to talk.

He asks, "What was that you said? That **midstov**? What's that?"

Mrs. Keane goes back to the bookmarked page and rereads the next to the last sentence, carefully enunciating: "**In the midst** of the winter, there are no flowers." She pauses and then asks: "Does anyone want to say what **in the midst of** means?"

Suddenly, the phrase catches on like a song, and the first graders are repeating it over and over. "In the midst of . . . midst of . . . in the midst of . . . midst of" They love it. It sounds so romantic and beautiful, so adult, too. It rolls off the tongue. It's a single unit, not four separate words, or three, or two.

But what does it mean? Oh yes, back to that. It is pretty much like **in the middle of**, according to Mrs. Keane. But the first graders have their doubts. **In the middle of** sounds so boring and stodgy. It can't possibly be anything like **in the midst of**.

Mrs. Keane rereads again. She gives them more examples. They trust her, to an extent. Rather grudgingly, they give in about what it means. She asks them to try using it in new sentences. They are really a lot happier saying **in the midst of** than talking about what it means. But they seem to be using it the right way, so Mrs. Keane is happy. Rest time begins.

A few days later the class is beginning a math lesson. Mrs. Keane takes out her math planning book and asks the kids to tell her which math lab they are working on so that she can make sure her records are up to date. Each math lab has a number. So the kids say things like "I finished lab 6 and I'm going to start 7." Or "I'm doing 4." Or "I just started 8."

When it's his turn, Tyrone grins and says, "I'm midstov 6."

Mrs. Keane smiles, too. So he's got the meaning, but he thinks it's just one word.

in this area. All classroom teachers should know how language development supports children's literacy development. They should know how to build on children's life experiences and general knowledge to help them comprehend what they read. They should be prepared with an array of techniques to bring the processes of reading comprehension to the surface so that they can demonstrate these processes and teach children to use them on their own.

Teachers with good preparation can make a world of difference for children—the difference between just calling out the words on a page and discovering the meaning, ideas, and life that can come from books and print.

She repeats the phrase **in the midst of**, separating each of the words as she writes slowly in her plan book.

The next boy says, "I'm almost midstov 7."

After a moment, Mrs. Keane catches on and offers a paraphrase: "So you've begun lab 7, but you haven't done very much of it yet." Maybe ignoring the pseudo-word will work.

"I'm midstov 4 and 5," Rachel says.

Okay, this girl has finished 4 and plans to start 5. "You're betwixt and between." "Mrs. Keane dangles other words to take the place of **midstov**, but no way, the kids don't bite.

Danny says simply. "I'm finished midstov 4." He hates math and doesn't want to talk about doing lab 5.

"I'll be midstov 6," Lakeisha says, meaning she plans to start lab 6.

And so on.

What a wonderfully silly class, Mrs. Keane thinks to herself. They get new vocabulary words and take them over. Talk about making a word your own. They don't want to give it back. They are wearing it out, stretching it beyond belief. It'll take some effort to get that genie back into the bottle! But she'd rather have a class that wants to use new words than a class that thinks vocabulary is a boring chore.

Mrs. Keane takes a second to tell them she's glad they like the words **in the midst of**. She emphasizes the **s** on **words**. She promises that tomorrow she's going to help them write **in the midst of**, and they'll put the words on their bulletin board and talk some more about it then.

As she puts away her plan book, her mind races: Maybe if the kids write the four separate words, they'll see. And then how to show them **in the midst of** is not the same as **between**. Most of the vocabulary work has been on single words. Wasn't there an article in last month's journal about first graders and word boundaries? Maybe Adele, the other first-grade teacher, can help plan a good vocabulary unit on phrases. Or maybe it just will not fit into the curriculum for first graders.

But now, on to the math lesson! Got to figure out about Danny hating math.

spoken language—a link to literacy

Classroom teachers must have a chance to pay more attention to spoken language as a foundation for written language. From the preschool years through the late elementary grades, a child's language development is a critical component of learning to read and write.

From infancy, children should be immersed in rich language experiences in their homes and child care environments. Too often though, early childhood educators are not sure how to engage children in interesting talk. Children need daily opportunities to witness adult models of language in use. Children need to practice their own conversational skills.

In the early years, by listening and talking, children learn the myriad connections and distinctions in the language—the *dog house* is not the same as a *house dog*; *Janie hit John* is quite different from *John hit Janie*; a tulip is a flower, but some flowers are not tulips; if something is not good, it is probably bad or close to it; if it's not red, it could be yellow or blue or green or white. Children gradually learn all the complexities. As the advertisement says, they just do it; they don't have to know its morphology and syntax, semantic hierarchies, antonyms, or semantic fields.

Without really thinking about it, children attach deep knowledge to words. Later, when they learn to identify words in print, they exploit that knowledge. Those connections and distinctions in language will help them figure out what it all means.

Many little ones arrive at kindergarten with a language repertoire that pertains mainly to the concrete, the "here and now." In books and in classroom discussions, they have to deal with a radically different situation and language that is sometimes quite abstract.

Language goes beyond the world around us. It brings a new world into existence; it gives a different viewpoint on the world. In school, children encounter descriptions of what happened long ago, what is happening in distant places, what could happen in the future, what might have happened if things were different in the past. As they move on in education (and abstraction), they have to be able to jump from questions about where a *nose* is on a *face* to ones about whether *justice* is a part of *democracy.*

Teacher education must ensure that teachers know the facts of language acquisition in early childhood and that teachers learn to help children manage the transition between spoken and written language. (See Chapter 1, pp. 37-38, for more about the distinctions between spoken and written language; see Chapter 3 for the relationship between printed words and spoken words.)

Often for young children, written stories, poems, and essays are delivered through a grown-up's spoken language. Well-prepared teachers learn to read aloud

in a compelling way, with clarity and style that will allure, entertain, and hold children's attention. One good way to pick up techniques is to watch and learn from expert storytellers like those featured on *Between the Lions,* a PBS television program.

Teachers learn skills for talking with children and for helping them talk with each other about what has been read to them—making connections, expecting explanations, looking for a punch line, asking about words that are new or used in an unusual way, noticing when they need to stop and think about what the book is all about. Teachers have to develop a repertoire of techniques to engage children as active listeners during a read-aloud.

Well-prepared teachers show children that whatever is spoken can be written, too. In lessons they elicit and write down children's comments. They set aside time in the day for whole stories, reports, and letters to be written, dictated to adults, written in collaboration with an adult, or written independently. And, always, the resulting products are read back and used, not just hung up like decorations or trophies.

Even when children become more independent readers, speaking and listening are still important parts of reading development. Teachers need to learn innovative ways to lead interactive discussions about reading. It is often a good discussion about the ideas, events, and facts that were read that really cements a student's understanding—or points out the need to read more in order to gain clarity and certainty. This is especially challenging when children's interests and tasks mean that they will not all be reading the same book—and may be reading books teachers have not yet read themselves.

Teachers must also stay alert in case a gap or stall in spoken-language growth becomes a barrier to reading as students encounter increasingly sophisticated materials. Teachers need to be able to use formal and informal methods for observing and assessing their students' language development, strengths, and needs. To get a complete picture, teachers should attend to how children use language during lessons and conversations with schoolmates, but they should also know enough about formal testing and clinical language assessments to turn to them as necessary. (See also Chapter 5.)

speaking of and in multilingual classrooms

In schools all over the United States the student body includes diverse language and cultural backgrounds. If they check, teachers might find three or four different languages spoken among families they work with and several different dialects of English, too. In some schools, teachers can expect even more diversity.

On the first day of school a child might find that one classmate has a Southern drawl and another speaks a northern inner-city vernacular—both dialects of American English. In the same class, one child might be terrific with both Spanish and English while another might be proficient in Hmong but at the beginning of the year needs help saying much beyond "Hi" in English.

Are language and dialect differences assets or problems? The quality of teacher education and professional development can be the deciding factor.

Parents, children, and teachers agree that children should become expert with mainstream American English. With teachers who treat standard English as an addition, not a replacement, classroom lessons can proceed smoothly and propel children toward the shared goal.

If teachers are well prepared, they know about the dialects of English and the languages spoken in the neighborhoods their students come from. They recognize when a different culture and language exert an influence in class discussions or in reading and writing. They are aware that many children have a language that works well for them outside school. They expect each child to become expert in mainstream English *through* school, finding it a useful addition for reading and writing and speaking in her or his brilliant future. They know how to help children use the language patterns in mainstream English comfortably and automatically, so that words pop out without conscious effort.

Colleges of education and school districts must ensure that teachers are prepared to find out about the languages, dialects, literacy, and cultures present in the areas

TEACHING JOSEFINA

Josefina writes We sat on Mary's porch and enjoyed. And no more. What happened? Does she want to use a different word than enjoyed but lacks the vocabulary? Is she tired, distracted, lazy—not finishing her sentence?

Mainstream American English use of the verb enjoy calls for an object after the verb (transitive verb is the official term). We expect the sentence to indicate something more about enjoy—like enjoyed ice cream, enjoyed the afternoon, enjoyed ourselves. No problem if the sentence had used an intransitive verb, like We sat on Mary's porch and relaxed. But that isn't quite the same meaning. We've all taken measures to relax in the midst of a stressful situation when we wouldn't claim to be enjoying ourselves.

But in Josefina's neighborhood a lot of people, like her family, learned the English dialect spoken in the Philippines. In that dialect the verb enjoy is intransitive, like the verb relax is in mainstream English. It's not a coincidence either that in Cebuano—the language that Josefina's parents often speak at home—the literal translation for enjoy is also intransitive.

Well-prepared teachers who work in Josefina's neighborhood have had a chance to learn about English dialects influenced by the languages of the Philippines. They know that Josefina's use of enjoy without an object isn't just due to vocabulary limits or distractions. The well-prepared teacher is ready to proceed without taking over or ignoring the message that Josefina intends to express. A discussion with Josefina can help figure out which word should be added. A well-planned lesson that focused on verbs and their objects can be planned, too. Over time, the teacher's job is to help Josefina develop the capability to handle mainstream American English and its use of transitive and intransitive verbs as well as many other grammatical structures.

It's not so different from the case of a British child who would say I went to hospital rather than I went to the hospital—and many other cases of different dialects of English.

where they teach. This information is basic for teachers to support the language development of students who are newcomers to speaking English. For these children the well-prepared teacher uses virtually every learning experience to practice, expand, and exercise their new language.

Children can (and perhaps must) do the two things at once—learn language and learn school subjects. While they are studying science—say, the fact that birds are a class of animals with special forms and behaviors—children are introduced to or practice words and sentence structures that are useful for more than discussions about birds. Whether English is a native language or a new language a child can find that the science lesson is a chance for language growth. But it is a *vital* opportunity for those learning English; more of the words and more of the language structures are new or only marginally known.

When the lesson takes place in a park or zoo, or otherwise calls for firsthand experiences, a lot of language learning can happen. But in classroom lessons children can make meaningful connections between the world and English, too, and teachers must be prepared to backstop and elaborate class discussions with multiple representations—different ways to introduce concepts and language and many chances to practice the new bits. Teachers must be able to recognize the words and language structures they will need to elaborate with children learning a new language.

Well-prepared teachers take the time and make the efforts that increase the amount and quality of language their students use. They say "tell me more" instead of a quick "good" before moving on to another topic or student. They paraphrase a child's answer, both to check if they understood the student and to model a more elaborate and grammatical expression of the idea. They respond with an enthusiastic "Really?" and stand still for a minute, letting a pregnant pause develop, so that the child will elaborate on the ideas he is eager to express. Teachers need to learn many tactics for helping students get the most out of the language they have and to get more language in the process.

Some students may need special support and instruction beyond the scope of their classroom teacher. Just because a child may quickly become good at speaking English to peers on the playground does not mean he or she is ready to rely on English for reading and learning academic subjects. Teachers need to learn to work with more than casual assessments and work to get children the help that is needed. A child having a good experience in a welcoming class is no substitute for work with a specialist prepared to teach English as a second or foreign language. (See the Resources section at the end of this chapter for recent information on working with speakers of languages other than English.)

vocabulary words— and the ideas that go with them

The more vocabulary words children learn, the better they can read and understand what they've read. If a text is full of unknown and therefore meaningless words, the reader will falter, become confused, and even give up.

It is important for teachers to have lots of tricks up their sleeves for teaching kids to learn new words and make it fun and interesting in the process. It simply doesn't work to send children to the dictionary to "look it up" and let it go at that.

Words do not lead quiet lives trapped between the covers of a dictionary. Words are vibrant things, born of experiences, places, nature, events, science, the body, food, inventions—in short, everything. When teachers give children new experiences, they get new concepts, and the words that go with them. A trip to a farm brings *tractor, hoe, plow, dairy.* Cooking offers *blend, broil, mushrooms.* A taste of a lemon ensures *tart.* An apple falling to the ground brings *gravity.* Play acting with other children clarifies words like *rescue.*

Through doing—class trips, cooking, art projects, science experiments, and role play—teachers can build children's vocabulary. Well-prepared teachers know how to capitalize on these experiences. They weave the new with what is already known. They teach children how to call on their existing knowledge to tackle new words.

They might rely on a semantic web—a definition in a map form that depicts word relations, including superordinate and subordinate categories as well as near synonyms and antonyms. A web begins with just a few words, sketched on the board, as the teacher has the students brainstorm about the word. For days, even weeks, the class may add to the web or readjust the elements in it, reflecting increasing familiarity and sophistication with the word and the semantic domain as a whole.

Word learning is not just a single-day activity that then gets discarded. Children learn particular words, but they also learn a bigger lesson—how to add new words to their vocabulary.

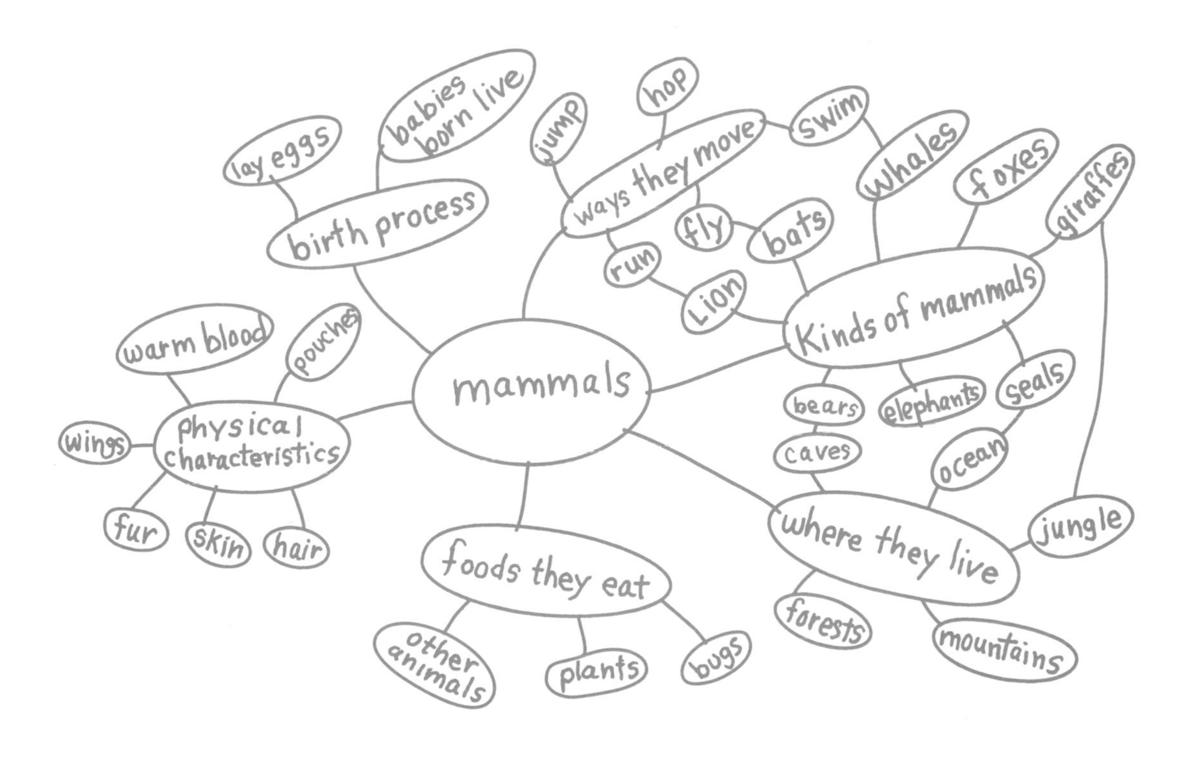

Joan Gottesman was a lawyer but then entered a teacher education master's program in New York City and interned at an elementary school.

I think that people who are good readers read in an organic, seamless way. When we read magazine or newspaper articles, we don't stop to say to ourselves, "Okay, now I'm going to read the heading to frame in my mind what this piece will be about and then I'm going to check back when I'm finished to see if my idea was right and then, if it wasn't, I'm going to try to figure out where the disconnect occurred and I'll go back and read it again and if I was right, I'll continue reading." But children who are struggling or who are not yet good readers need to become conscious of their reading habits so that they can monitor what they do and what they don't do and learn the comprehension strategies that good readers use unconsciously, so that they, too, can achieve the kind of automaticity that characterizes good readers.

Understand Before You Move On

When I studied the teacher's role in reading comprehension in my master's program, I had some of those "ah-ha" moments, where suddenly (or so it seemed) I realized that although we know that reading has two components, decoding and comprehension, we rarely focus on the deliberate part of teaching comprehension. I think there are a lot of students who don't exactly understand that reading is both parts, that decoding without comprehension is not reading—it's just decoding—and that good readers make sure they understand the meaning of their texts before they move on to the next clause, sentence, or paragraph. That's okay, because we can help these learners become good readers by explicitly teaching them how to comprehend a text.

My courses helped me visualize a framework for explicitly teaching comprehension. Having an outline in my mind was almost like having index cards with the types of strategies written on them, that I could shuffle and pull out as I needed them. It's not as if there are only two prereading, two during-reading, and two postreading strategies that a teacher can use, but thinking about categorizing my knowledge in a grid or framework makes it easier

Learning vocabulary through reading is important, too. Effective teachers immerse children in a particular subject to help them build vocabulary. For example, the teacher might read *Charlotte's Web*, by E. B. White, to a class of first graders. This is a book that most first graders cannot read on their own but nonetheless will find enrapturing. To go with it, the teacher finds materials on spiders that the children can read independently—poems or nature books, perhaps. Children get multiple chances to develop deeper knowledge of new words when they find them in different contexts and with different styles of writing.

Well-prepared teachers also learn to give their students "word awareness." This means giving kids the drive, zest, and playful desire to learn new words because they know words are fun and valuable. Without word awareness, students are more likely to skip over words they don't know and jeopardize comprehension. The teacher needs to let students know that not knowing at first is okay. Figuring out the meaning is what everybody does.

for me to think about consciously teaching it. You have to scrutinize your texts to analyze what demands they place on the readers, in terms of vocabulary, inferencing, or other comprehension tasks. Matching readers to texts is a real art and takes time but is worth it because you can build so much more from strength than from weakness.

This is what I learned about being a teacher for children's comprehension growth: You have to think about stuff, go back and deconstruct your own processes, and then examine those of your students. What are their strengths? Where are they falling down? Do they understand what comprehension is?

I learned that when teaching comprehension strategies you model what you're trying to teach. You tell students that a technique you're presenting is a strategy, a way to approach something, and often one that they can use in a variety of situations. Let's say a student noticed that there were paragraph headings. You could discuss what they are and why students think they are there. If no one thought they were there for organization, you'd suggest reading a paragraph together and then go back and see if there was a connection between the text and the heading. You'd ask an explicit question: "Does anybody see a connection between the heading and what we just read? Show us." If no one saw it, you'd point it out. You'd talk about what people noticed and then try it again and again. You'd maybe ask a question like "So how do you think you could use paragraph headings to help you with your reading next time?" or "Can you think of another kind of reading where using the heading might help you understand what you're reading?" That would probably be enough for one day (or too much for one day).

Model it, and practice it, practice it, practice it.

When you teach comprehension, you are teaching students how to think, how to make connections, and how to think about their thinking. I don't ever want to teach students what to think. I want them to be able to understand what I think—and what an author thinks—and to decide for themselves what they think.

Teachers must learn that creative approaches to words help children take on word awareness. For instance, the teacher might have a routine for the class to work on mystery words every now and then. The teacher explains that a mystery word is one that the children can figure out how to say once they see it written—very simple spelling-sound correspondences. But the words should be ones the children are unlikely to have met before—out of context the children would be stymied about a word's meaning. First, the teacher makes deputies of the students so that they are ready to track down the meaning and provide the evidence that will stand up in the court of time and memory! She holds up the victim—the word written alone—*clambers.* The students say it easily. Next the teacher shows the group a short written passage, bordered in yellow with crime scene tape, the mystery word outlined in chalk: *The boy clambers up the tree. He puts the egg in the nest.* The group of deputies will quickly suggest "climb" for a meaning, and the teacher will press for the evidence and the reasoning, calling attention to "up" and "the nest" and what everyone knows about trees and birds' nests. In case there is need for more hints, the teacher might have another exhibit that includes the passage on a page with an illustration of a boy awkwardly climbing a tree. Next, to narrow it down—the difference between *climb* and *clamber*—a dictionary or thesaurus comes into the lesson. The lesson concludes with the deputies using the word in new sentences of their own making. The children could learn the vocabulary item in many different ways; the well-prepared teacher recognizes that this way has the bonus of promoting a class climate of word awareness.

An excellent teacher knows how to promote word awareness by creating interest in homographs, words that are spelled the same way but have quite different meanings. This is another recurring opportunity to teach children that readers pay attention to how words are used in sentences to figure out meaning. A teacher might set up a poster with a thermometer like the ones used for fund raising, but for this one the units are indicated by a word that has two distinct meanings and the red column increases each time the class finds one of these homographs. The teacher should be alert to make contributions, too. For instance, most kids know that the word "lying" means not telling the truth. But in the book *Whistle for Willie,* what does it mean when the text says "He hid in an empty carton *lying* on the sidewalk"? Teachers should be able to help their students figure out the "other" meaning based on the passage, and then one more homograph can be added to the thermometer poster.

comprehension and metacognition

Comprehension is sometimes misunderstood as a fairly passive experience of "remembering" what we have read or adequately "receiving" the ideas of the author. But comprehending is a very active process. In order to remember anything, in order to receive anything, we have to roll up our sleeves and get to work.

We have to pull up background knowledge and ideas to begin making sense. Then we have to jettison some of our old ideas, or at least remodel them, as we learn from the text. We contribute our reading skills and strategies to the job. We bring our own psychological affect to the interpretations. We have a certain purpose that makes some parts of what we read more important than others. With all this we create and re-create until we get an understanding that works. Then it's simple—"That's just what it says," we claim, overlooking all that we did to *make* it say something.

Nowadays, most classroom teachers have had a chance to take courses that cover concepts important for teaching comprehension—the role of background knowledge in understanding; inferences; specific strategies like summarizing, predicting, and questioning; methods for direct instruction; and applications of

strategies in reciprocal teaching and other kinds of instructional conversations. (See the Resources section for studies of comprehension and its instruction.)

Too often, though, these concepts are kept in the abstract; they are not applied to specific passages or only in a limited way, perhaps only to fictional narrative prose passages. The concepts and techniques are often demonstrated one at a time, as if any passage called for only one strategy to be used—by the teacher or the student reader.

The real world of understanding is a lot busier. We do it every time we read something, but noticing "real-world" understanding—the intellectual processes within it—is difficult. The trick is to catch understanding processes while we are reading! When we are aware of these intellectual processes, we are *metacognitive.* That means we can think about our own thinking as we read.

What good is that? Doesn't it just add clutter and take the reader off task? Yes, it's true that metacognition is not very useful when everything is going well. But that's just the point. Children limit their achievement if they read only what is easy, where they can be sure "everything will go well." When they read material that stretches them, that makes room for improvement, then metacognition is useful. Children need metacognitive strategies to fix reading that goes awry, in fact, to be able to notice that it *is* going awry.

Teachers, like other accomplished readers, do not find it "natural" to read and think about reading at the same time. Too many teachers have not had a chance to step back and notice their own comprehension strategies. But if teachers are not aware of their own comprehension strategies, they miss a gold mine of resources for teaching children. When a teacher has learned to talk explicitly about how one goes about reading a passage, in effect a window opens into the teacher's head and children can peek in and see how it's done.

Well-prepared teachers have had a chance to develop three capacities: to be metacognitive about their own reading, to bring the hidden processes to the surface for their students to see, and finally to help children manipulate their own hidden processes when it will help them to get over problems and read more challenging material more independently.

Even teachers who are adept at metacognition need special practice to bring the hidden processes to the surface when a passage is easy for them to read. It isn't easy for children, so it is worth it for teachers to demonstrate how they read, how they check their understanding along the way, and how they make repairs when they need to. Well-prepared teachers can read and keep up a running commentary, asking themselves questions and answering them, making skilled reading visible for the children to witness and learn from.

This kind of performance is called a "think-aloud." A think-aloud should cover preparations for reading, from the mundane and general (good light, good chair, some protection from distractions) to specific attention to the passage. Teachers give children copies of the passage so they can follow along. While scanning titles and headings and pictures, noticing a boldface or italic word here and there, the teacher comments on what she already knows and what she'll probably be able to find out, what might be easy and what might take more effort.

A think-aloud combines reading aloud and interrupting the reading. The teacher pauses to make a comment about something she wants to remember, explaining what gave a hint that it might come in handy for later reading. Another pause comes when something doesn't make sense. The teacher rereads something, explaining what put her on notice that there was a problem, how she decided to try rereading to solve the problem, and how she knew how far back to go.

The think-aloud continues after the text has all been covered. It can include a summary "in your own words," an application of the ideas in the passage, a question the passage brought up, a goal to do some more reading, or some other activity.

Teachers need an opportunity to practice thinking aloud and working with students to follow up on them. There should be time for children to ask about the teacher's think-aloud and time for the teacher to ask the students if they ever do what she does when she reads. It's the time to delve deeply and explain directly some comprehension strategy that the class is ready to learn but hasn't been relying on enough.

Effective teachers are prepared to follow up with careful attention to children when they are reading on their own. That's the time to remind them that comprehension takes work for everybody—it's not magic; it's a matter of working at it. That's the time to remind them about a particular strategy demonstrated in the teacher's think-aloud and to help them apply it to problems they are having right then and there. At first, teachers and students can use the strategies together, but the teacher's role fades out.

The well-prepared teacher keeps track of the techniques and strategies the children are able to use and prompts them to use them throughout the classroom day—reading for different school subjects, reading for different purposes, always reading for comprehension.

Activity

Words in the World

We are usually as unaware of language demands as we are of a clean window when looking through one at a beautiful garden. Teachers must learn to look at the window—at the language—if they are to help their students conquer it. In this activity, teachers have a chance to notice the complexity of experiences from which new concepts and words are built. Beyond empathy for the children's tasks, they also develop materials for some specific lessons. The facilitator for this set of lessons should have a background about teaching students who are new to the language of instruction.

1. Divide the class of teachers or teachers-to-be into teams of three. In an urban neighborhood, tell each team to visit a city block's worth of restaurants. Each team's task is to look for all the repetitive features noticed in each restaurant. Remind the team members to listen, too, for repeated things that people say. No notes. Just remembering.

   - Back in the classroom, list each team's findings on the board. Check the other teams' lists and go back and add words each team forgot or missed but "knows" were repeated features of the restaurants it visited.

   - First, discuss words that people hadn't known before, or knew but weren't sure if they knew how to read or write, or knew only with a more limited meaning. Then consider features they took so much for granted that they didn't list them at first. Have the class consider the different aspects of knowing a word.

   - Finally, discuss special cases—the features that are the same in the abstract but a little bit different each time. One example is: "I'll have a coke/Pepsi/cup of coffee." It's really a repeated language structure with a variable slot. How hard is it to talk about words that are collections not single instances? How much harder are truly abstract words like "cost"?

- Discuss how to prepare children to read or listen to a story set in a restaurant. How can teachers be prepared to help new English speakers handle the language demands like the ones the group has been discussing?

- In a suburban setting, base the exercise on the similar and different features of houses on a street, the architecture, the landscaping, and what people say and do when they answer the door.

2. For homework tell the teachers to repeat what was done in class, but one-fourth take trips to art museums and one-fourth to science centers, while the other half prepares to use new printed matter, some a new story and the rest a new science book.

3. The next class is the time to pool lists and then ideas for being ready to support language growth for students. At least four advice packets should be produced. After the class reviews and corrects them, make copies for the whole class of the most useful ideas. Tell the class to be prepared in a few weeks to revisit the issue, reporting on an idea the teachers tried and how they know if it worked.

Activity

What Does That Word Mean Here or Anywhere?

The following exercise helps teachers become aware of their own strategies for figuring out the meanings of words, first with a simple word that has special meaning in context and next with odd words that most teachers or teachers-to-be will not have met before.

1. Make the arrangements needed to photocopy the first page of two novels: *Beloved* by Toni Morrison (Plume/Penguin Books, 1988) and *Jane and the Wandering Eye: Being the Third Jane Austen Mystery* by Stephanie Barron (Bantam Paperbacks, 1998).

2. Write "124" on the board. Have students say what it means—maybe 12 tens plus 4, maybe 1 more that 123, etc.

3. Write this sentence on the board: "124 was spiteful." (Tell those who remember Tony Morrison's novel to zip their lips, please.) Discuss how hard it is to figure out what the sentence means because few if any of the meanings given for 124 give a hint that it could be spiteful. Some may claim the sentence is meaningless.

4. Pass out the photocopies of the first page of *Beloved*. Give a few minutes for everyone to read it. Discuss the meanings: "124" is a house number, a word to refer to a house, the house that characters in the novel lived in, a spirit, a child's spirit, etc.

5. How do you know what the word means here?

   - Discuss specifically which words or phrases on the page gave the teachers information about the meaning of "124." Go sentence by sentence through each paragraph.
   - Give names to the kinds of context clues that recur here and that the teachers know are useful for reading other passages, too. Put aside the list of context clue types and the page from *Beloved*.

6. Ask the class members to try to define five words without using a dictionary. The words are *rout-party, efflorescence, negus, pasties,* and

stupid. Someone from the United Kingdom might get *pasties;* a master of prefixes, suffixes, and roots might get *efflorescence. Stupid* will seem easy. There may be creative responses for the other words.

7. Pass out a photocopy of the first page of *Jane and the Wandering Eye,* a fictitious diary with a mystery twist, alleged to have been written by Jane Austen in 1804. Give everyone a few minutes to read it. This is the opening paragraph:

   "A rout-party, when depicted by a pen more accomplished than my own, is invariably a stupid affair of some two or three hundred souls pressed elbow-to-elbow in the drawing-rooms of the great. Such an efflorescence of powder shaken from noble wigs! Such a crush of silk! And what general heartiness of laughter and exclamation—so that the gentler tones of one's more subdued companions must be raised to a persistent roar, rendering most of the party voiceless by dawn, with only the insipid delights of indifferent negus and faltering meat pastics as recompense for all one's trials."

8. Discuss what *stupid* means here. Consider how the word refers as much to a speaker's opinion as to the noun it is supposedly modifying.

9. Next turn to *rout-party, efflorescence, negus,* and *pasties.* Repeat step 5 above. Point out that this time the clues are used for words that readers didn't know before, not just meanings special to a context. Ask which word they are least sure of. Suppose a dictionary could be used to check only one of the words. Which one do they vote to look up? (*Negus* is the least supported by clues in the text. The reader can infer that it is a kind of drink, but. . . .)

10. Discuss the lists of context clues. Ask for reports of experiences with context clues for meaning, for both unknown words and words used in unusual ways. Provide the class with references about context clues and instruction. Assign the teachers to work in teams to plan a lesson that will help children use context clues to understand a passage and be able to use them subsequently on their own. In later class sessions, study and critique the plans and, if, possible, the results of using them.

Activity

Getting Metacognitive with Gloopy and Blit

This is a way to help teachers become aware of the intellectual processes involved in reading. The trick is to use a strange text. The reader has to work overtime to make sense of it. All that effort shines a spotlight on the background knowledge, procedures, and strategies that are tried and that are discarded or that succeed. The reader's ordinarily hidden intellectual work becomes noticeable when the writing is so strange. It's almost impossible to fail to be metacognitive. It is a good introduction to readings about metacognitive strategies in comprehension instruction.

Two-thirds of the teachers or teachers-to-be pair up to work on the following selection. Each pair is joined by a classmate who will take notes on what happens while they work.

Gloopy and Blit

Gloopy is a borp. Blit is a lof. Gloopy klums like Blit does. Gloopy and Blit are floms.

Ril had poved Blit to a jonfy. But he had not poved Gloopy.

"The jonfy is for lofs," Blit bofd to Gloopy, "you are a borp."

Gloopy was not klorpy. Then Blit was not klorpy.

- The pairs are to figure out whatever they can about the words, sentences, paragraphs, and the whole passage. They should have some prompts to use if they get stuck: Read it. Summarize it. Are there main characters? Are there any other characters? Compare and contrast each character. List a few things that happened in chronological order. Did anything cause anything else? What questions would you predict a "next" chapter would answer? Substitute words and come up with a passage that makes sense. Write it down.

- The observer should have some prompts too. Do they read aloud? When and how often do they reread, aloud or to themselves? Do they say why? Can they pronounce all the words? Do they mention letter-sound correspondences or analogies to pronunciations of known words? Do they complain about the number of unknown words? Did anyone doubt for a minute that Gloopy, Blit, and Ril were names? Did anyone comment on this and say how they knew? Do they use the "new" words to summarize or answer any of the "comprehension" questions? Or do they actually substitute words before they do the other tasks? Do they mention any features of writing, like uppercase letters, punctuation, or paragraphing? Do they mention any features of morphology, such as which words are nouns and which are verbs and how they know that? Do they mention anything about syntax or sentence structure? Did they suggest a category—a semantic interpretation—to capture the relationship between the main characters? Did they rely on typical narrative structure? How did their prior experiences come into play? Were some overwhelmed by the nonsense language and others fascinated by the challenge of the task?

After 15 minutes, have the trios discuss the experience, using the observer notes and prompts as a guide, adding or correcting by using the metacognitive insights from the reading pair. Conclude by picking two ideas about reading that came up and one question about reading and prepare to present them to the class.

After 30 minutes, convene the class as a whole. Ask for a volunteer to read a story that resulted from substituting words in the passage. Ask about similar and different stories that resulted. Most likely no one substituted words for Gloopy and Blit, so shift the conversation to the metacognitive by asking about that: What is it about names as a special case of words we don't know? Then ask for volunteers to present their "two ideas" about reading. Discuss them, and note who else had the same ideas. Cover the questions in a similar way.

For an assignment, provide readings about metacognition, noting that a reader who does what all three of the members of the trio did is being metacognitive.

In a subsequent class, model reading a short passage while thinking aloud about the reading. Reassemble the trios, and have each student choose a passage from a children's book to practice solo metacognitive think-alouds, helping each other to notice and report on what they do to comprehend.

resources

Basic reviews of the knowledge base relevant to this chapter can be found in the following recent publications:

Kamil, M. L., Mosenthal, P. B., Pearson, P. D., & Barr, R. (Eds.). 2000. *Handbook of Reading Research: Volume III.* Mahwah, NJ: Erlbaum. (See especially Nagy and Scott on vocabulary, Alexander and Jetton on learning from text, Goldman and Rakestraw on structure and meaning, Pressley on the essence of comprehension instruction, Blachowicz and Fisher on vocabulary instruction, Wade and Moje on varying texts, Bean on reading in different subjects, Bernhardt on second-language reading, Garcia on bilingual reading, and Au with a multicultural perspective on equity and excellence.)

National Reading Panel. 2000. *Teaching Children to Read: An Evidence-Based Assessment of the Scientific Research Literature on Reading and Its Implications for Reading Instruction: Report of the Subgroups.* Washington, DC: National Institute of Child Health and Human Development. (See especially Chapter 4, Part I, "Vocabulary Instruction," and Part II, "Text Comprehension Instruction.")

National Research Council. 1998. *Preventing Reading Difficulties in Young Children.* Committee on the Prevention of Reading Difficulties in Young Children, C. E. Snow, M. S. Burns, and P. Griffin, Eds. Washington, DC: National Academy Press. (See especially Part I, Chapter 2, and Part III.)

Neuman, S. B., & Dickinson, D. K. (Eds.). 2001. *Handbook of Early Literacy Research.* New York: Guilford Press. (See especially Watson on literacy and oral language, Scarborough on aspects of early language development related to subsequent reading difficulties, Bus on storybook reading, Morrow and Gambrell on literature-based instruction, and Tabors and Snow on young bilingual children and early literacy development.)

The sections of this chapter can be elaborated by consulting various other sources, including the following:

Language development:

Berko-Gleason, J. (Ed.). 1997. *The Development of Language, 4th Edition.* New York: Allyn & Bacon.

Dickinson, D. K., & Tabors, P. O. (Eds.). 2001. *Beginning Literacy with Language: Young Children Learning at Home and School.* Baltimore: Brookes.

Fillmore, L. W., & Snow, C. E. 2000. *What Teachers Need to Know About Language.* Washington, DC: Center for Applied Linguistics.

Hart, B., & Risley, T. R. 1999. *The Social World of Children Learning to Talk.* Baltimore: Brookes.

Snow, C. 1983. Language and literacy: Relationships during the preschool years. *Harvard Educational Review, 53,* 165-189.

Snow, C. 1991. The theoretical basis for relationships between language and literacy in development. *Journal of Research in Childhood Education, 6,* 5-10.

Snow, C., & Goldfield, B. A. 1983. Turn the page please: Situation-specific language acquisition. *Journal of Child Language, 10,* 551-569.

Speakers of languages other than English, general issues, reviews, and overviews:

August, D., & Hakuta, K. (Eds.). 1997. *Improving Schooling for Language-Minority Children: A Research Agenda.* Washington, DC: National Academy Press.

Bamford, J. D., & Richard, R. 1998. Teaching reading. *Annual Review of Applied Linguistics, 18,* 124-141.

Bialystok, E. (Ed.). 1991. *Language Processes in Bilingual Children.* New York: Cambridge University Press.

Coady, J., & Huckin, T. (Eds.). 1997. *Second Language Vocabulary Acquisition.* New York: Cambridge University Press.

Fitzgerald, J. 1995. English-as-a-second-language reading instruction in the United States: A research review. *Journal of Reading Behavior, 27*(2), 115-152.

González, V. (Ed.). 1999. *Language and Cognitive Development in Second Language Learning.* Boston: Allyn & Bacon.

Greene, J. P. 1997. A meta-analysis of the Rossell and Baker review of bilingual education research. *Bilingual Research Journal, 21* (2/3). Available online at *brj.asu.edu/archives/23v21/articles/art1.html.*

Reyes, M. L., & Halcón, J. J. (Eds.). 2001. *The Best for Our Children: Critical Perspectives on Literacy for Latino Students.* New York: Teachers College Press.

Tinajero, J. V., & Ada, A. F. (Eds.). 1993. *The Power of Two Languages: Literacy and Biliteracy for Spanish Speaking Students.* New York: Macmillan/McGraw-Hill.

Wiley, T. G. 1996. *Literacy and Language Diversity in the United States.* McHenry, IL: Center for Applied Linguistics & Delta Systems.

Speakers of languages other than English, samples of more specific topics and studies:

Au, K. H. 1998. Social constructivism and the school literacy learning of students of diverse backgrounds. *Journal of Literacy Research, 30*(2), 297-319.

Boers, F. 2000. Metaphor awareness and vocabulary retention. *Applied-Linguistics, 21*(4), 553-571.

Crowell, C. G. 1995. Documenting the strengths of bilingual readers. *Primary Voices K-6, 3*(4), 32-38.

Eviatar, Z., & Ibrahim, R. 2000. Bilingual is as bilingual does: Metalinguistic abilities of Arabic-speaking children. *Applied Psycholinguistics, 21*(4), 451-471.

Field, M. 1996. Pragmatic issues related to reading comprehension questions: A case study from a Latino bilingual classroom. *Issues in Applied Linguistics, 7*(2), 209-224.

Fitzgerald, J. 1999. About hopes, aspirations, and uncertainty: First grade English-language learners' emergent reading. *Journal of Literacy Research, 31*(2), 133-182.

Garcia, G. E. 1991. Factors influencing the English reading test performance of Spanish-speaking Hispanic children. *Reading Research Quarterly, 26*(4), 371-392.

García, G. E., & Nagy, W. E. 1993. Latino students' concept of cognates. *National Reading Conference Yearbook, 42,* 367-373.

Goldenberg, C. 1994. Promoting early literacy development among Spanish-speaking children. Pp. 171-199 in E. H. Hiebert & B. M. Taylor (Eds.), *Getting Reading Right from the Start: Effective Early Literacy Interventions.* Boston: Allyn & Bacon.

Jimenez, R. T., Garcia, G. E., & Pearson, P. D. 1995. Three children, two languages, and strategic reading: Case studies in bilingual/monolingual reading. *American Educational Research Journal, 32*(1), 67-97.

Jimenez, R. T., Garcia, G. E., & Pearson, P. D. 1996. Three reading strategies of bilingual Latina/o students who are successful English readers: Opportunities and obstacles. *Reading Research Quarterly, 31*(1), 90-112.

Lee, J., & Schallert, D. L. 1997. The relative contribution of L2 language proficiency and L1 reading ability to L2 reading performance. *TESOL Quarterly, 31*(4), 713-739.

Mace-Matluck, B. J., Hoover, W. A., & Calfee, R. C. 1989. Teaching reading to bilingual children: A longitudinal study of teaching and learning in the early grades. *NABE Journal, 13,* 187-216.

Moll, L. C., & Gonzalez, N. 1994. Lessons from research with language minority children. *Journal of Reading Behavior, 26*(4), 439-456.

Muñiz-Swicegood, M. 1994. The effects of metacognitive reading strategy training on the reading performance and student reading analysis strategies of third grade bilingual students. *Bilingual Research Journal, 18*(1-2), 83-97.

Nagy, W., García, G. E., Durgunoglu, A., & Hancin-Bhatt, B. 1993. Spanish-English bilingual students' use of cognates in English reading. *Journal of Reading Behavior, 25*(3), 241-259.

Truscott, D. M. 1997. Supporting cultural and linguistic diversity in beginning readers. *The New England Reading Association Journal, 33,* 21-25.

Ulanoff, S. H., & Pucci, S. L. 1999. Learning words from books: The effects of read-aloud on second language vocabulary acquisition. *Bilingual Research Journal, 23*(4), 409-422.

Vocabulary and concepts, sample of reports and studies:

Beck, I. L., & McKeown, M. G. 1991. Conditions of vocabulary acquisition. Pp. 789-914 in R. Barr, M. Kamil, P. Mosenthal, and P. D. Pearson (Eds.), *Handbook of Reading Research, Vol. II.* White Plains, NY: Longman.

Carlisle, J. F. 1995. Morphological awareness and early reading achievement. Pp.189-209 in L. B. Feldman (Ed.), *Morphological Aspects of Language Processing.* Hillsdale, NJ: Erlbaum.

Fischer, U. 1994. Learning words from context and dictionaries: An experimental comparison. *Applied Psycholinguistics, 15*(4), 551-574.

Fukkink, R. G., & de Glopper, K. 1998. Effects of instruction in deriving word meaning from context: A meta-analysis. *Review of Educational Research, 68*(4), 450-469.

Kuhn, M., & Stahl, S. 1998. Teaching children to learn word meanings from context: A synthesis and some questions. *Journal of Literacy Research,* 30, 119-138.

Leung, C., & Pikulski, J. J. 1990. Incidental word learning of kindergarten and first grade children through repeated read aloud events. In J. Zutell & S. McCormick (Eds.), *Literacy, Theory and Research. Analyses from Multiple Paradigms.* Chicago: National Reading Conference.

McKeown, M. G. 1993. Creating effective definitions for young word learners. *Reading Research Quarterly, 28*(1), 16-31.

McKeown, M. G., Beck, I. L., Omanson, R. C., & Pople, M. T. 1985. Some effects of the nature and frequency of vocabulary instruction on the knowledge and use of words. *Reading Research Quarterly, 20*(5), 522-535.

Nagy, W. E., & Scott, J. A. 1990. Word schemas: Expectations about the form and meaning of new words. *Cognition & Instruction, 7*(2), 105-127.

Robbins, C., & Ehri, L. C. 1994. Reading storybooks to kindergartners helps them learn new vocabulary words. *Journal of Educational Psychology, 86*(1), 54-64.

Scott, J. A., & Nagy, W. E. 1997. Understanding the definitions of unfamiliar verbs. *Reading Research Quarterly, 32*(2), 184-200.

Senechal, M., & Cornell, E. H. 1993. Vocabulary acquisition through shared learning experiences. *Reading Research Quarterly, 28*(4), 360-374.

Stahl, S. A. 1998. *Vocabulary Development.* Cambridge, MA: Brookline Books.

Waring, R. 1997. The negative effects of learning words in semantic sets: A replication. *System, 25*(2), 261-274.

Comprehension and metacognition:

Alvermann, D. E., & Swafford, J. 1989. Do content area strategies have a research base? *Journal of Reading, 32*, 388-394.

Applebee, A. N. 1978. *The child's concept of story: Age two to seventeen.* Chicago: University of Chicago Press.

Armbruster, B. B., & Armstrong, J. O. 1993. Locating information in text: A focus on children in the elementary grades. *Contemporary Educational Psychology, 18*(2), 139-161.

Beck, I. L., McKeown, M. G., Hamilton, R. L., & Kucan, L. 1997. *Questioning the Author: An Approach for Enhancing Student Engagement with Text.* Newark, DE: International Reading Association.

Brown, A. L. 1997. Transforming schools into communities of thinking and learning about serious matters. *American Psychologist, 52*, 399-414.

Brown, A. L., & Day, J. D. 1983. Macrorules for summarizing texts: The development of expertise. *Journal of Verbal Learning and Verbal Behavior, 22*, 1-14.

Brown, R., Pressley, M., Van Meter, P., & Schuder, T. 1996. A quasi-experimental validation of transactional strategies instruction with low-achieving second-grade readers. *Journal of Educational Psychology, 88*(1), 18-37.

Cross, D. R., & Paris, S. G. 1988. Developmental and instructional analyses of children's metacognition and reading comprehension. *Journal of Educational Psychology, 80*(2), 131-142.

Duffy, G. G., Roehler, L. R., & Herrmann, B. A. 1988. Modeling mental processes helps poor readers become strategic readers. *The Reading Teacher, 41*, 762-767.

Fountas, I. C., & Pinnell, G. S. 1996. *Guided Reading: Good First Teaching for All Children.* Portsmouth, NH: Heinemann.

Gambrell, L., & Almasi, J. F. 1996. *Lively Discussions! Fostering Engaged Reading.* Newark, DE: International Reading Association.

Gambrell, L., & Jawitz, P. B. 1993. Mental imagery text illustrations and children's story comprehension and recall. *Reading Research Quarterly, 28*, 264-273.

Guthrie, J. T., & McCann, A. D. 1996. Idea circles: Peer collaborations for conceptual learning. Pp. 87-105 in L. B. Gambrell & J. F. Almasi (Eds.), *Lively Discussions! Fostering Engaged Reading.* Newark, DE: International Reading Association.

Guthrie, J. T., Van Meter, P., Hancock, G. R., Alao, S., Anderson, E., & McCann, A. 1998. Does concept oriented reading instruction increase strategy use and conceptual learning from text? *Journal of Educational Psychology, 90*(2), 261-278.

Keats, E. J. 1964. *Whistle for Willie.* New York: Viking Press.

Markman, E. M. 1979. Realizing that you don't understand: Elementary school children's awareness of inconsistencies. *Child Development, 50,* 643-655.

McDermott, R., & Varenne, H. 1995. Culture as disability. *Anthropology & Education Quarterly, 26,* 324-348.

Palincsar, A. S., & Brown, A. 1984. Reciprocal teaching of comprehension fostering and comprehension-monitoring activities. *Cognition and Instruction, 1,* 117-175.

Paris, S. G., Cross, D. R., & Lipson, M. Y. 1984. Informed strategies for learning: A program to improve children's reading awareness and comprehension. *Journal of Educational Psychology, 76,* 1239-1252.

Pittelman, S. D., Heimlich, J. E., Berglund, R. L., & French, M. P. 1991. *Semantic Feature Analysis: Classroom Applications.* Newark, DE: International Reading Association.

Pressley, M., & Wharton-McDonald, R. 1997. Skilled comprehension and its development through instruction. *School Psychology Review, 26,* 448-467.

Pressley, M., El-Dinary, P. B., Gaskins, I. W., Schuder, T., Bergman, J. L., Almasi, J., & Brown, R. 1992. Beyond direct explanation: Transactional instruction of reading comprehension strategies. *The Elementary School Journal, 92,* 513-555.

Raphael, T. E., & Hiebert, E. H. 1996. *Creating an Integrated Approach to Literacy Instruction.* Orlando, FL: Harcourt Brace.

Rosenshine, B., & Meister, C. 1994. Reciprocal teaching: A review of the research. *Review of Educational Research, 64,* 479-530.

Rosenshine, B., Meister, C., & Chapman, S. 1996. Teaching students to generate questions: A review of the intervention studies. *Review of Educational Research, 66*(2), 181-221.

Schoenbach, R., Greenleaf, C., Cziko, C., & Hurwitz, L. 1999. *Reading for Understanding: A Guide to Improving Reading in Middle and High School Classrooms.* San Francisco: Jossey-Bass.

Snow, C. E. 2002. *Reading for Understanding: Toward a Research and Development Program in Reading Comprehension.* Santa Monica, CA: RAND.

Strickland, D. S. 1991. Emerging literacy: How young children learn to read. In B. Persky and L. H. Golubchick (Eds.), *Early Childhood Education, 2nd ed.* Lanham, MD: University Press of America.

Westby, C. 1999. Assessing and facilitating text comprehension problems. In H. W. Catts and A. G. Kamhi (Eds.), *Language and Reading Disabilities.* Boston: Allyn & Bacon.

White, E. B. 1952. *Charlotte's Web.* New York: Harper Collins.

Whitehurst, C. J., Falco, F. L., Lonigan, C. J., Fischel, J. E., DeBaryshe, D. B., Valdez-Menchaca, M. C. & Caulfield, M. 1988. Accelerating language development through picture book reading. *Developmental Psychology,* 24:552-559.

Sounds, Letters, and Words

How Print Works

3

The 26 letters are the obvious part of reading. But what do those letters represent? Thinking about the letters alone is as odd as thinking about the "2 + 2" side of an equation and not the "= 4" side. Written English is based on an *alphabetic principle:* Letters "stand for"—symbolize, represent, correspond to—units of spoken language.

The alphabetic principle is not the only possible design base for a writing system, but it is the way that English, Spanish, Vietnamese, Russian, and many other writing systems work. In these systems, letters are associated with small units of sounds called phonemes. The alphabetic principle does not prescribe one unique letter for each unique sound though. There are plenty of cases in English of words with spellings we can complain about. But even as we complain, we take for granted the alphabetic principle and take issue only with the consistency or directness of its use. For English the alphabetic principle is applied with a *deep orthography.* Orthography means spelling. A deep orthography goes beyond phonemes. The H in GHOST shows that etymology—the history of language structures—is in our orthography. The same letter C for different sounds in MAGIC and MAGICIAN demonstrates the effect of morphology—the forms and different categories of words.

The alphabetic principle raises questions about the form and structure of language. Teachers need ample opportunity to study them. Otherwise, they are handicapped as they try to teach children in three of the areas of competence that are essential for learning to read an alphabetic system:

- **Phonemic awareness.** Beginners notice that a spoken word has parts that are unrelated to its meaning—just sounds that can be separated and combined, ready to be represented by letters in written words.

- **Phonics.** Readers use systematic correspondences between letters in written words and sounds in spoken words. Phonics drives the search of the crowded storehouse of memory to locate words that readers have met before. Phonics provides the material for readers to construct pronunciations of written words they never before met.

- **Fluency.** Fluent readers coordinate word identification and comprehension, achieving their purposes for reading, having put processes and strategies on "automatic pilot."

THE and THEN

Billy had an urge to read to Sandy Doan, the intern teacher. He wanted to show off how much he had learned, and he knew he could ask for help now and then. When Sandy helped, she applied what she knew about phonics and about Billy. She started at the beginning of the word and went left to right and blended, so that Billy could put the parts together and recognize the word. When the book had patterns that hadn't been the focus of lessons yet, Sandy was quick to fill in, but she made sure to let Billy's new learning shine through whenever it could be applied. Sandy knew that Billy had just finished studying a unit focusing on short *e* with words like BET, MEN, and LED.

The reading was going fine until they came to the word THEN. Two lines above Billy had read HEN with no hesitation. Sandy's plan was to start with TH, providing the sound herself if there was any hesitation but making sure to give Billy the floor for the rest of the word. She covered up the end of word, and together they pronounced a sound for the first two letters. Then she uncovered the vowel, and Billy took over and read confidently to the end of the word. "THUN," Billy said, beginning like THE but ending like SUN.

"THUN?" Sandy asked, confused as her plan went awry.

"TH, THE, THUN," Billy paused between the successively larger parts. "What's that?"

As soon as she heard Billy say THE, Sandy saw the trap she had fallen into. There was a bulletin board in the classroom with a short list of words. It was labeled "Instant Words." The class would go over the list, and the children and teacher would chant, "What do we do about instant words? We know them; we see them; we say them." The word THE was at the top of the list. And, no doubt, here were exactly the same letters in the word THEN!

Sandy learned from this. It was crucial for her to help Billy apply both sorts of word identification strategies—instant words and phonics patterns like short *e*. As a teacher she would have to be ready for "traps" that could come up when it looked like both strategies could apply.

Right now she said, "Oh, Billy, that's another instant word. It's THEN." Billy read it out and sailed on to the rest of the sentence.

Later Sandy had a lot of thinking and talking to do with her mentor teacher so that she could be prepared to avoid or handle other problems like this. Was there something in the teachers' manual about it? Maybe there was an article somewhere. Maybe it was just something to learn from Billy!

These three big ideas call for a significant time commitment for teachers' professional preparation, even though they take up a small proportion of the total class time for children. Teachers need broad and deep knowledge about sounds, letters, and words. They must apply that knowledge skillfully in different tasks with different children.

What do effective teachers do with such knowledge? While good materials and good curricula are available and helpful, it takes a well-prepared teacher to find them and enact them for the benefit of each child. Well-prepared teachers know how to examine a textbook to check its completeness. Suppose a phonics program covers final N-T and final N-D consonant blends (as in BENT and BEND) and covers S-T (as in BEST) but leaves out S-D. The well-prepared teacher is not worried about the gap, knowing the omission is not an error but a reflection of the structure of the English language.

Effective teachers take the time to be sure that the materials, learning tasks, and assessments do not trip up individual students or mislead the teacher about a child's progress and understanding. Suppose a child says that DON and DAWN sound the same, but the book says they are not, or a child treats TOLD, TOLL, and TOE as homophones, just like TOO, TWO, and TO, but the teacher recognizes differences in the endings of the words. A well-prepared teacher knows these are two predictable consequences of the diversity in American English, not necessarily wrong or inattentive responses to phonemic awareness lesson tasks. A teacher must use different sets of words to gauge children's phonemic awareness. Similarly, a young first grader may consistently read the TR cluster as if it were CH. The well-prepared teacher recognizes a common pattern of spoken-language development and does not confuse it with insufficient application of phonics.

Teachers are the not-to-be-missed link for children's learning and development. A teacher's knowledge about the facts of language is a crucial resource. Teacher preparation and professional development must help teachers apply this resource in their day-to-day practice regarding phonemic awareness, phonics, and fluency.

facts about sounds and spoken words

Too few teachers have had a chance to learn the facts about language. What parts does a language have? How do children develop language? What differences are there among languages and even within a single language? Specialized courses about the teaching of reading should rest on a solid bedrock of knowledge about language.

Medical schools expect entering students to know biology; it is a prerequisite. Law schools expect their candidates to know about the history and structure of government.

Schools or departments of education should be able to count on incoming students knowing a good deal about language. Liberal arts and science departments should ensure that service courses teach about the structure of language and the language development of children.

Nowadays, though, it is seldom the case that entering education students are so prepared. So teacher education curricula and professional development plans must take the time and develop the resources to teach the facts of language.

There are several layers of language structure that teachers need to understand. Two are especially important for dealing with sounds and words—morphology and phonology. Neither term is in the common parlance. We begin with morphology. Though an unfamiliar term, it involves familiar concepts.

Morphology and Morphemes

Past tense, plural, root, suffix—these are familiar from foreign-language or writing classes. Such categories are part of the morphology of any language. Morphology is the study of the smallest meaning-bearing units in language. Morphemes are these units—words and meaningful parts of words.

The word FRIENDS has two morphemes: {friend} is the root morpheme and {s} is the plural morpheme in the suffix. In FRIENDLY the suffix changes the class of the word; it turns a noun into an adjective. The prefix in UNFRIENDLY changes the meaning to the opposite. BEFRIENDED uses the root with a prefix that turns it into a verb and a suffix that can signal past tense. UNBEFRIENDED is going too far, but we can push through to a meaning if we absolutely must.

Morphological patterns do *not* replace phonics patterns; they augment them. Morpheme patterns with high frequency can serve as guides to break words into "chunks." It is too hard to read very long words letter by letter, left to right. Proficient readers can rely on manageable chunks made of roots and affixes (prefixes or suffixes). Perceiving the chunks in a word eases memory searching for more fluent identification of known words. Morpheme chunking also lets readers build up to words they haven't seen before. Sometimes the reader recognizes that the chunks add up to a word known from listening or speaking but that is new in the reading vocabulary. Other times the reader learns a completely new word, putting together the morphemes to construct the meaning at the same time that he or she meets the spelling and pronunciation.

There are two types of affixes in morphology that teachers have to be prepared to handle: derivational and inflectional. Derivational morphemes change the meaning of the root (FRIENDLY becomes UNFRIENDLY) or the class of the word (FRIEND, the noun, becomes the verb BEFRIEND). They appear as both suffixes and prefixes in English.

Sometimes the derivational morpheme affects the reading and pronunciation of the root. The last sound in the adjective ELECTRIC changes when {ity} is added for the noun ELECTRICITY but changes in a different way when {ian} makes the noun ELECTRICIAN. Sometimes spelling and pronunciation differences cover up the fact that the same root or affix is working in a pair of words. The root is affected in EVADE/EVASION or CONCLUDE/CONCLUSION. The prefix is affected in IMPERFECT/ INDUBITABLE. Morphemes like {in} that originate in Latin are notorious shape shifters, becoming IR in IRREVERENT, IM in IMMATERIAL, and so on.

The inflectional morpheme is the other type of morpheme that teachers need to know about. In English there are only eight inflectional morphemes, but they are used very often. They are all suffixes, and each one works with a particular class of words. Sometimes, the root affects the pronunciation of the suffix. An example is the {s} inflection for plural nouns. This plural suffix is read with the same sound in words like TRICKS and POTS but a different sound in a word like DOGS. The morpheme adjusts to the last sound of the root word.

Teachers should know that children learn many derivational morphemes while they learn to read. Inflectional morphemes, though, children do not need to learn in school. They should learn to *use* them in reading and writing, but they already have a complete inflectional system when they come to school. They may not know they know it, but they use it and take it for granted. A great deal of research shows that the various dialects of English do not skimp on inflectional morphology. Whatever dialect of English children bring from home, the teacher can rest assured that it works to distinguish between past and present, for instance, and singular and plural. A home dialect may mark number on a noun or tense on a verb in a different way than happens in the system the teacher takes for granted, but the child (and community) knows and expresses the distinctions.

Teachers need to know enough about English morphology to use it while teaching reading and, if need be, to adapt materials for lessons with the particular students in their classes. They need to know how morphology can work in other languages, especially languages like Spanish with which so many American school children have early experience. Teachers need to be prepared to work effectively with children who find the English morphological patterns so unusual that they stumble over them rather than find them helpful for word identification during reading.

THE SCIENCE BEHIND PHONEMES AND PHONOLOGY

How do scientists find the phonemes of a language? How do they know a phoneme when they see one? Well, actually, they have to see more than one at a time to see just one. Linguists investigate how phonemes function in a stream and how they contrast in a system. Phonemes function in the words of a language; they function to indicate differences between words. The contrasts among all the phonemes form a system for the language.

The word "lit" and the word "let" each have three phonemes—three chances to be different. Cutting the words up into any smaller or larger pieces won't work in the same way. The first phonemes—/l/ in phonemic notation—are the same in the two words, no matter if they seem a little different to a machine or a speaker of another language. Substituting the first phoneme from "lit" for the one in "let" will not make a different word. The same goes for the third phoneme, /t/.

But switch the second phonemes—/ɪ/ and /ɛ/—and you've switched words. This substitution procedure with minimal pairs is a tool for identifying phonemes. If "lit" and "lip" are the minimal pair, we see that the third phonemes make the difference; if we investigate "lit" and "bit," the functioning of the first phoneme is highlighted.

"I lit the candle" means something quite different from "I bit the candle," and the phoneme that signals the meaning difference has nothing to do with matches or teeth! Switch one phoneme and switch the whole meaning, even though neither phoneme alone means what the word it is in means! The phoneme makes the difference but doesn't mean the difference.

So what does /l/ mean? A good definition of /l/ is the following: /l/ is not /ɪ/ or /ɛ/ or /t/ or /b/ or /p/ or /k/ or /n/, /s/, /r/, and so on for all the other phonemes of English. It's an odd definition but not unique in our experience. Like the definition of "middle child" depends on contrasts among ages of siblings in a family, what a phoneme means is also a matter of contrasts but within the phonology of a language.

Phonology and Phonemes

Knowing about phonology is important for teachers of beginning readers because the basic units—phonemes—are what letters of the alphabet represent.

Examples are a good way to start thinking about phonemes. The spoken words MOUSE and HOUSE are alike except for the first phoneme. Change the middle phoneme of PEG and it turns into PIG. Leave out the second phoneme in SPAT and it's SAT. MAID, FADE, and PLAYED rhyme—the same phonemes at the end, no matter the spellings. But PLAID does not have the same middle phoneme as MAID, no matter the spelling, and, of course, they don't rhyme!

Children learn phonemes coincidentally as they acquire language to negotiate

XYZ

While phonemes have no meaning outside this system of contrasts, we can describe them acoustically, that is, by diagrams of the sound waves they are typically associated with. Without instruments, we can also describe phonemes in terms of the way they are typically made. Called articulatory descriptions, these cover the flow or obstruction of the breath stream through the mouth or nose and the movement or position of the tongue, lips, and vocal cords. Take "bit" as an example:

With /b/ the two lips clamp together to stop the flow of breath while the vocal cord vibrates—technically, a bilabial voiced stop;

With /ɪ/ the front part of the tongue is in a high position, toward the roof of the mouth, without much tension in the tongue muscle and the air stream flows freely—technically, a lax high front vowel;

With /t/ the tip of the tongue touches the roof of the mouth behind the teeth to stop the air flow and the vocal cords don't vibrate—technically, an alveolar voiceless stop.

Studies of the phonology of English and other languages are an active area of science. Many questions are still to be resolved. A current example of interest for beginning reading involves entities bigger than phonemes and smaller than syllables. Linguists specializing in phonology argue and provide evidence for a three-unit syllable subdivision—onset, nucleus, coda. For the syllable "strict," /str/ is the onset, /ɪ/ is the nucleus, and /kt/ is the coda. Psychologists and educators sometimes use a two-unit subdivision—onset and rime. /str/ is the onset and /ɪkt/ the rime.

Should linguists move to a two-unit division? Should they keep the three units but add the two units as another layer in phonology? Should psychologists and educators move to a three-unit subdivision? Now is the time to see what precise questions to ask, what evidence to assemble. Tune in to the further exploits of phonologists for the answers—and new phonology questions.

their way through childhood, getting cookies and hugs, not pulling the dog's tail, and trying to put off bedtime for a little while longer. People use phonemes without being aware of their existence. People rely on all aspects of phonology without thinking about it.

In fact, native-language phonology gets to be so much a part of us that it can undermine attempts to learn another language. No matter how we try to fix our mouths, those two French words come out the same even though the French teacher tells us they have to be different. We cannot believe our ears when the Spanish teacher tells us that two sounds are alike when they sound so different to us!

And this is the crux of the matter: Ears are not enough. Hearing phonemes involves the phonology of a language, not just human ears. Phonemes are categories—categories that are part of the phonology of a specific language.

In English, PA and SPA have the same /p/ phoneme. A machine measuring movements of the mouth and the breath stream would indicate considerable difference between the first sound of PA and the middle sound of SPA. So would a graph of the resulting sound waves. In another language, like Korean, those kinds of differences are categorized as two different phonemes. But it is just one phoneme in English. The phonology of English differs from the phonology of Korean, or Spanish, or, in fact, all other languages.

The phonology of a language is not just a list of phonemes. Phonology defines categories like vowels and consonants. It defines syllables and permitted patterns of phoneme combinations within a syllable.

In American English, for example, the phonemes /mps/ are permitted to appear together in that order as a consonant cluster. Think about words like CAMPS or JUMPS. But there are not and never will be words or syllables in American English that *begin* with /mps/ or even with the smaller consonant cluster /mp/. This is a finding from the study of the phonology of American English.

When English "borrows" a word from another language, the word is often adjusted to the phonology of English. If the borrowed word starts with /mp/, we patch it up to fit into our phonological system. We add an unaccented vowel before the /m/ or between the /m/ and the /p/. There is a poet from Malawi whose name is difficult for English speakers to pronounce: Edison Mpina. Listen for the

patch you use to handle the cluster forbidden in American English! We may borrow, but we keep our interest in English phonology!

People who speak English as a second language often show systematic traces of the phonology of their native language. A grandmother talking about a FILM says what sounds like FILL 'EM to her granddaughter. Nana's first-language phonology does not allow /lm/ consonant clusters. Similarly, SPECIAL becomes ESPECIAL for those whose first-language phonology forbids initial /sp/ clusters. The patching strategies that bring other phonologies into accents for American English are like those that occur when words are borrowed into English from other phonologies.

Teachers need to be ready for the individual differences in language development they will find in their classrooms. They should be able to recognize ordinary developmental pacing and to notice language- or speech-related problems that should be referred to a specialist. By age 6, most children speak and understand the essence of the phonological system used in their community. Of course, some aspects are consolidated earlier than others. For example, for quite a few kindergarten children, the word JOKE is pronounced like DOKE, but within the next year the initial sound of JOKE becomes a reliable part of their language production.

A final important set of facts for teachers to know about phonology is how it reflects the language diversity in the United States. Each teacher should know the features of the dialect she or he speaks. All teachers should also know about

ABC ENGLISH IN THE UNITED STATES—NO KING OR KING'S ENGLISH

Some Americans say the following pairs of words are pronounced the same:

- WHALE and WAIL
- PIN and PEN
- SOD and SIDE
- SOD and SAWED
- FULL and FOOL
- POOR and POUR
- DOOR and DOUR

Some of us agree on one pair but disagree about another. Who's correct? There's no way to tell. We do not have a national language academy.

When the words in any of these pairs are claimed to be the same, two phonemes have been merged. Some of these mergers have been spreading in the United States for more than 100 years; others are recent. Mergers can be unconditional—applying to all words that have the phoneme at issue—or they can be limited to words with certain phonology or morphology patterns, or even limited to an exceptional word or two.

Like other aspects of language change and variation, a merger can be found in some regions of the country and not others, in cities and not in the countryside, or vice versa. A merger can be more evident among women or among men, more widespread among one age group than another.

"Chain shifts" are another way that English dialects show diversity in the United States. The following is a contemporary vowel shift in cities from Syracuse to Chicago:

- the vowel in CAD moves to the position of the stressed vowel in IDEA;
- COD shifts and sounds like CAD to people from other regions;
- CAWED shifts and sounds like COD;
- KED shifts to sound like CUD;
- KID shifts but just sounds odd, not like a different phoneme of English to others.

Besides differences in the inventory of phonemes, languages and dialects have different conditions for syllable structures. English allows "closed" syllables. BEE is open and BEET, ending with a consonant, is closed. But there is variety here, too.

the variation of English and other languages spoken in the communities they work in and about the influence of contact with other languages spoken in the community as well.

Language differences are inevitable over time. While literature, film, and drama alert us to the difference between Shakespearean English and our everyday language, we seldom notice that the phonology of American English from the year

XYZ

In the northern United States, JIMMY expects that the first syllable of his name is JIM, a closed syllable, while the second syllable, my, is open. If he goes South, he hears his name with an accent—it's JI (open) and MY (also open). He may perceive the accent as charming or annoying, but it's just a part of life in a diverse country.

English permits clusters of two or three consonants at the end of syllables, like at the end of test. Final consonant clusters are often simplified—one consonant is deleted.

The deletion occurs more in some dialects of English than in others, and the conditions under which consonants are deleted differ from dialect to dialect. Both phonology and morphology get involved. If the final consonant signals a past tense (a morphology matter), the process of simplification is inhibited in African American vernacular English.

Compare what happens with TEST and MESSED. Remember, for many of us they rhyme—despite the spelling, they have the same three last phonemes. But MESSED has the verb plus a past tense verb morpheme at the end, while TEST is just one morpheme for the whole word. In the African American vernacular English dialect, the one morpheme TEST is much more likely to be simplified—pronounced without the final /t/—than is the two-morpheme word MESSED. So this pair of words would not be a good example of a rhyme!

Some people decry the diversity of American English as evidence of decay; others romanticize any differences. Sociolinguists actively study the diversity in English in the United States and know it shouldn't be oversimplified in either direction. Current research addresses a range of issues: Who are the harbingers of change? How does it spread? When is diversity evidence of change in progress? What aspects of language are ripe for diversity or change? Which are universal aspects of diversity and change and which are specific to a certain language or society? They ask, too, how ways to teach reading and prepare teachers can best be informed by results from sociolinguistic studies.

For detailed information on language diversity in the United States see Atlas of North American English: Phonetics, Phonology and Sound Change, by W. Labov, S. Ash, and C. Boberg (New York: Mouton de Gruyter, 2002).

1800 was different than the phonology of 2000. There is even evidence of differences between generations. Geography and society matter, too, in language diversity—the region of the country, urban or rural roots, ethnic background, gender, and class relations. In places where two languages have coexisted over time, like English and Spanish in the Southwest, dialects of both languages reflect the language contact—even for people who speak only one of the languages.

penetrating words—phonemic awareness

To speak or understand a language we depend on phonemes, but we do not need phonemic *awareness* to speak or understand. Children in kindergarten and beginning first grade are proficient at speaking and listening; they know how to use phonemes. But they may not know about what they know.

As we learn to read an alphabetic writing system we become aware of the sounds inside words enough to talk and think about them. For beginning readers and writers there is a crucial "ah-hah" experience—the recognition that each letter in a written word can be related to a part of a spoken word. But suppose the beginner is not aware that there are little parts, phonemes, in spoken words. Suppose, that is, that the beginner is not phonemically aware?

Before age 5 or so, children typically do not have a way to talk about phonemes or to acknowledge their existence in any other way—to recognize or produce them on purpose, for instance. In a discussion about the sounds in the word CAT, they can contribute lively comments about meows and purrs, but they'll be mystified if the conversation focuses on sounds of consonants and a vowel in a certain sequence.

Separating the word from its referent is not an easy task for very young children; they do not delve into the phonemes that make up the spoken word. Children are attentive to syllable structure at a young age, before they are sensitive to the boundaries between words and phonemes. To observe this, just ask preschoolers or young kindergarten children to repeat lines from the Pledge of Allegiance. In the last line, some children will say "liver tea" for the word "liberty" in the original. The children preserve the structure of the syllables, but the words and phonemes are not as salient. It should not surprise the teacher that very young children are better at counting syllables than they are at counting words or phonemes.

Phonemic awareness and reading are mutually reinforcing. It's not a one-way street. So why not save everyone's time and effort by focusing on reading and letting phonemic awareness take care of itself?

There are several reasons to spend teacher preparation time on phonemic awareness. Each is based on evidence from numerous research programs. First, children who display phonemic awareness are more likely to be good readers at a later age. Second, instruction focused on phonemic awareness works well with children who display no initial awareness, even before they have learned to recognize letters reliably. Third, children who become phonemically aware as a result of such programs later perform better while learning to read words than do similar children who have not participated in phonemic awareness instruction.

What is it that a phonemically aware person knows? He or she knows that spoken language has a form—a spoken word can be broken into parts. He or she

Ruth Nathan, Ph.D., third-grade teacher, Rancho Romero Elementary School, Alamo, California

They Simply Didn't Understand What Was Being Asked

I got my first teaching credential in my mid- 20s. I got a job in Des Moines, Iowa, teaching upper elementary. I had a number of children, even back then, who couldn't read, didn't have a clue. And I didn't have a clue how to teach them. I had a full undergraduate teaching credential but no decent course to teach reading. I took time off to raise my children. Later, I went for a master's degree in reading.

Now I am a third-grade classroom teacher, and I teach graduate classes at a nearby university. First-grade and kindergarten teachers need to understand phonemic awareness, but most of them don't. They don't understand that kids need to move from the meaning of words to the idea that words are sounds that can be broken into different parts.

I never fully understood this myself until I was asked to lead in-service training of teachers. I knew a lot about the alphabetic principle and the importance of teaching phonics, but I never totally understood phonemic awareness.

So with a few colleagues, I put myself back in a graduate school mode. We met, read, visited kindergarten classrooms, and did research ourselves, until we got it. We learned how to do diagnostic phonemic awareness tests and went out and assessed kids.

Then the lightbulb went off.

I realized why some kids don't understand phonics. It's because when their teachers were instructing them, the students weren't phonemically aware. They didn't know how to take a word and break it up into sounds. It wasn't necessarily because they **couldn't** do it, but they simply didn't understand what was being asked of them. I learned that phonemic awareness could be taught, and ordinary classroom teachers can teach it. And then kids can rely on it to know what they are supposed to do in phonics lessons.

knows that a language has relatively few sounds that are recycled, rearranged, for use in many different words. Take CAP, HAT, and CAT. Five of the phonemes of English reappear among the nine phonemes that make up these words. Knowing that ordering matters is another part of phonemic awareness. SPOT is not the same as POTS even though the words use the same phoneme segments.

In sum, phonemically aware people know not only their language but also about their language. *Metalinguistic* is the general term for this kind of knowledge. It is really quite a trick. Humans do a great deal that they are unable to describe or otherwise display awareness about. If you try to say explicitly what you do to go up a flight of stairs, you can see the problem!

What tasks show whether someone is phonemically aware or not? There are essentially three basic types of activities. The first two have to do with the phonemes in a single syllable. *Synthesis* activities start with parts and make a whole. A sample task is: "Given the phonemes /m/, /ɛ/, and /n/, blend them to pronounce a word." The answer is *men. Analytic* activities do the opposite. A sample task is: "Given the word *men,* report how many phonemes it has." Three is the answer.

The third kind of phonemic awareness activity has to do with the phonemes in the system of the language, not in one syllable or word at a time. *Identity* activities locate phonemes that are the same but that occur in different words and in different places within words. A rather tricky identity task follows: Take the eight words MET, NEW, MINE, TOM, NEAT, TEND, PAN, SMALL. Sort them into two sets, one set of words that uses the phoneme /m/ and the other with /n/. The answer is one set with four words and one with five words. Notice that MINE works in both sets—the first phoneme is identified as /m/ and the last phoneme as /n/.

In kindergarten Nate wrote a four-sentence story about a favorite toy. He could read the story to his teacher after he wrote it and even a week later. "This is a eyeball. It has veins. The pupil is black. The eyeball rolls."

What Does Writing Have to Do with Phonemic Awareness?

Very young children's writing attempts are often more like drawing or decoration, so phonemic awareness is not relevant. When scribbles and letter-like forms enter the picture and especially when recognizable letters appear, awareness of phonemes gets to be virtually inevitable. In writing there are opportunities for all three types of phonemic awareness activities—analysis, synthesis, and identity. The child *analyzes* the sounds in a message in order to choose letters to represent it in writing. When the young author reads back the message, there is an opportunity

for *synthesis*. Phoneme *identity* comes into play as a young author looks to models or written words to choose letters for a sound or word.

A child's first spellings are full of interesting problem solving. As the child provides estimated spellings, the alert teacher can observe and, when appropriate, assist the child's attempts to attend to all the phonemic segments in a word. Typically, beginners use letters to represent only the beginning sound of a syllable. They may combine the lone letter with scribbles or lines indicating the relative length of the word.

In time, children try to represent all the sounds in all the syllables of a word. They estimate which letter to use when. They often notice a similarity between a sound in the word they want to write and the sound of the letter name. Nate, for example, spells VEIN with an A as the vowel in the middle. Once a child begins to spell conventionally, using models from spelling lessons and reading experience, memory searches and spelling principles overshadow problem solving.

Teachers must be prepared to amplify children's early writing attempts. They can offer tools and advice to spotlight phonemic awareness as children are writing on their own or to dictation. In one dictation practice, sometimes called "writing for sounds," the teacher stretches out the sounds in each word, enunciating in an exaggerated way, modeling part of the analytic work on phonemes that she is promoting among the children.

Some have expressed fear that children who estimate spellings in early writing will later fail to learn conventional spellings. There is no evidence to substantiate that fear.

The best spellers among us still estimate spellings each and every time they look up a word to see how it is spelled. They have to estimate the spelling of a word—using hints from phonemic analysis and conventional spelling in tandem—to find it in the dictionary! Of course, we defer to the dictionary's conventional spelling and abandon our estimation. Not only is it the graceful thing to do, it's practical if we want our writing to communicate to others!

Nate kept writing and reading back, branching out into narrative fantasies: "I sa a Algadr and He smild At Mo and we wat nito the wdr," read back as "I saw a alligator and he smiled at me and we went into the water."

Hannah writes (and later reads back): My parents love me no matter what. They give me presents on my birthday and on Christmas day.

Countering myths

Phonemic awareness instruction is newly popular, and some widely circulated information and assumptions are erroneous. During their professional development, teachers should be put on guard about myths, such as the following:

- **Myth 1:** *Phonological awareness and phonemic awareness are the same thing.* No, some manipulations of the sounds of language are *not* phonemic awareness tasks. A child may be asked to segment sentences into words, or words into syllables, or syllables into onsets and rhymes. These are phonological awareness activities. They may be useful as an introduction for phonemic awareness tasks but are not themselves at the phoneme level, the level that matters for the alphabetic principle.

- Myth 2: *Auditory discrimination is the same as phonemic awareness.* No, phonemic awareness tasks must maintain a metalinguistic element. Say a child recognizes that the third word is different in the set BIT, BIT, BEAT. This is not a display of phonemic awareness. Perceiving the difference is a part of linguistic, not metalinguistic, knowledge.
- Myth 3: *Phonemic awareness curricula teach children the phonemes of a language.* No, children begin acquiring the phonemes of their native language in infancy and essentially have finished by kindergarten. Phonemic awareness activities do *not* replace programs for children learning English as a second language or programs that address hearing problems or speech and language delays or disorders.
- Myth 4: *All the phonemes of the language must be covered in a phonemic awareness curriculum.* No, there is no need to cover all the phonemes of a language. Good effects have been achieved with programs that use just a few consonants and vowels that can be combined to make many syllables and words. No specific set of phonemes has been shown to be crucial for complete instruction. More and more time on phonemic awareness activities does not produce better and better results.
- Myth 5: *All children need the same instruction in phonemic awareness.* No, research has not shown that there is a single best task or type of phonemic awareness activity. Nor has it shown that a particular sequence of phonemic awareness tasks is required. In any given kindergarten or first grade, some students will be well beyond the need for any phonemic awareness experiences and will make the best progress by getting on with their reading. Other children in the same class may not be able to respond to the phonemic awareness tasks without special preparation covering one-to-one identity or sorting tasks.
- Myth 6: *Phonemic awareness tasks involve only oral language, never letters.* No, while the focus is on the spoken word, there are ways to make phonemic awareness tasks more concrete. Children can manipulate tokens, like blocks or tiles, while they synthesize, analyze, or identify phonemes in spoken words. Letters need not be avoided in phonemic awareness instruction. In fact, research has shown that it is helpful to use tiles with letters that correspond to one or more of the phonemes in the task.

corresponding with letters—phonics

Proficient readers may not be aware of all the ins and outs of the phonics connecting letters and sounds for reading English. They may not be able to talk about phonics patterns, but their actions show their proficiency. A reader uses phonics:

- to quickly and accurately search memory to identify words met in print before;
- to figure out a pronunciation for a never-before-seen printed word and to search for its identity among words known through speaking and listening; and
- to identify the sounds that go with a printed word and begin learning it as a new word through reading.

Phonics has a long history in the lore of teaching reading and teacher preparation. Every teacher of the early grades knows something about it, so we emphasize the "news." The key word in contemporary discussions of phonics is *systematic.*

A good way to understand what systematic phonics means is through a simple word like CAP. Think about the following part of the system that includes CAP. CAP is an example of a word pattern that comes early and often in books for beginners: the *consonant + short vowel + consonant* pattern. But you do not really know CAP unless you know that CAPE does not have the sound of CAP in it. CAPE is an example of another equally important pattern for beginners: *long vowel + consonant + e*. Sometimes this is called the "silent e" pattern. In some longer words the letters of CAP appear with a different vowel sound. Systematic patterns in multisyllable words allow us to predict the "lazy" vowel (schwa) in the unaccented syllable in ESCAPADE as well as another appearance of the long vowel in the first open syllable of CAPABLE.

Consider just the first segment of CAP. What does a C say? Or, better to ask, what does a C say at different times? Compare the C in CAP and PACK and with the T in TAP and PAT. The T works pretty much the same way at the start or the end of the word, but the C takes a K along to say its sound at word's end. In ACCURATE, two C's do the work (like the two T's in ATTITUDE). In CEMENT and CIGAR, the C says something else, sounding like an S often does. Words in a set like the one with CUT and KIT bring up the other consonant, K, that says the same thing C does in CAP. And sometimes they overlap and give us homonyms—CANE, KANE—though the second is a proper noun. In CHICKEN and CHOOSE, C is part of a consonant digraph and has yet a third sound, but beware of the CH in words from Greek like CHORUS because that gets back to the sound in CAP. And what if we add the IAN suffix to magic? The C sound is like SH.

No wonder we refer to the alphabet as the ABC's and stop at the C! The letter C provides a good idea of the complex but systematic phonics a proficient reader relies on for reading English.

What do children learn to become proficient? It's like the modern business method of just-in-time inventory. Children need to learn just enough to be able to start self-teaching for word identification. Everyone knows that school cannot teach children each and every word they need for reading—close to 100,000 for success in the elementary grades! School cannot possibly cover even the 500 or so patterns of letter-sound correspondence. Luckily, there is good evidence that self-teaching of phonics takes over once children have a good start. School must give children a chance to work successfully on some of the letter-sound patterns. School must give them a chance to practice applying the patterns they learn and a chance to practice self-teaching—finding, learning, and using new patterns.

Curricula vary about how phonics patterns are introduced and used. *Synthetic* methods begin with a focus on a letter (or letters) that represent a single phoneme and blend them into words or syllables. *Analytic* approaches begin with words and segment them into smaller letter-sound correspondence units. And there are mixed approaches that go from *part* to *whole* to *part* again. There are also programs that consider slightly larger units. They rely on patterns of onset and rime segments or groupings of words in word families. CAP yields C as onset and AP as rime. Substituting other onsets like T, L, and S produces TAP, LAP, and SAP for a word family. Consonant blends can be added—TR and STR for TRAP and STRAP.

Phonics programs of all types recognize that some words that occur very frequently in speech and writing are spelled in unexpected ways and need special treatment. For example, H-A-V-E does not spell a word that rhymes with SAVE, even though many words with A-V-E do rhyme with SAVE—like CAVE and PAVE. Note, too, the beginning of WHO does not sound at all like the beginning of WHAT despite the same first two letters. Frequently used words that are oddly spelled are learned as whole words, not analyzed or synthesized. They are often called "sight" or "instant" words.

Of course, the goal is to read all words at a glance, quickly, as if there is no underlying phonics process going on. When a word first enters a child's reading vocabulary, it betrays the process that was involved in learning it. It may show traces of segmenting and blending at the single-sound level or at the onset-rime level. Or it may show a trace of the analogy from known word families or the way it was memorized as an instant word. If a child is to comprehend and read for a variety of purposes, he or she must become an "instant reader" of just about everything so that meaning can be the main event.

The need for teacher expertise about phonics is great. On the one hand, phonics programs on their own are not enough. There is no evidence that there is a single

best phonics program that can be used off the shelf. There is no evidence about which letter-sound correspondence patterns must be included, which are the best methods of using them for word identification, or which sequence of instruction steps should be followed. A variety of approaches have produced good results, but if there are required ingredients, research has not yet found them.

On the other hand, incidental and occasional phonics instruction is not enough to prevent reading difficulties and start children out right. Most children profit from more than a mention of phonics patterns or occasional use of methods for identifying letter-sound correspondences to identify words. It's not enough for a teacher to be a good reader who occasionally shares her expertise at processing words as the occasions arise. There's too much in the system of letter-sound correspondences and it is too complex. There's little chance of finding all the crucial examples, of improvising explanations, and of providing sufficient practice on the spot.

It is up to teachers to choose and use good materials and the best-possible curriculum. Even if the teacher has a great program that worked beautifully for the children in last year's class, the teacher is responsible for ensuring that it or something else works to suit each child in this year's class. Teachers' knowledge of phonics underlies their ability to enact the curriculum under changing day-to-day conditions, to monitor its effectiveness, and to augment or repair it where needed. In-depth teacher knowledge is required to support children's practice with applying the knowledge of phonics patterns and strategies on materials outside the reading lesson. The informed teacher provides the safety net as children become self-teaching and venture into challenging materials. The teacher's deep knowledge of phonics guides his or her choice of materials and methods of assistance as the children attempt to apply the phonics in their reading for different purposes.

A well-prepared teacher's knowledge base includes the research about how children approach reading and word identification, even before there is any formal instruction. In order to support children's success in the most efficient ways, teachers must see how their plans, lessons, and observations are affected by the stages and phases that children go through on their way to full proficiency. Teachers must also have a good grasp of the content of phonics—the common letter-sound correspondences and the patterns found in multisyllabic words. Finally, teachers need to keep abreast of the research and practice about effective methods for teaching letter recognition and about the synthetic, analytic, or mixed approaches to teaching word identification. To keep up with the news and to make the best choices for their students, teachers need to know terms and techniques for a variety of approaches to phonics, even though they need not (and probably should not) use them all during any one instructional year. They must know about and be able to coordinate with the school-, district-, or state-mandated curriculum and materials.

fast, easy, complete reading—fluency

Fluent readers assemble the reading process in full. They identify letters, sounds, and words. They understand. They are quick, accurate, automatic. They read with intonation and tempo that suit the meaning. Among other things, fluent readers:

- emphasize meaningful contrasts with the number word in "SUSAN HAS TWO BROTHERS AND JENNY HAS *THREE* BROTHERS." For "SUSAN HAS TWO BROTHERS AND JENNY HAS TWO *SISTERS*," they emphasize the family relation contrast.
- end on a higher pitch for "DID YOU OPEN IT?" and "YOU OPENED IT?" There is a chill if they read "DID YOU OPEN IT?" with a falling intonation—a hint of blame or accusation. But fluent readers end the intonation pattern lower for "WHY DID YOU OPEN IT?" unless the meaning includes clarification or rhetorical effect: "WHY DID YOU OPEN IT? BECAUSE YOU ARE JUST LIKE PANDORA!"
- make use of the comma to pause and to make sense of "TO PATRICK, HENRY WAS A HERO."

The text helps or hinders fluency. Some material provokes fluency problems. Doubters should read a technical article for a profession they have not studied. There can be false starts, frequent stops, mispronunciations, and a need to reread almost every sentence and paragraph to understand the material. The more we read the same or related material, the more fluently we read.

A beginner might find that just about all texts present a fluency problem. Teachers need to know how to move a halting reader toward fluency. Fluency is the worry when we say, "Chris read each word correctly, but it seemed like she had no idea about the meaning," or "Alejandro sounded out almost every word. It was so slow and disjointed that no one could possibly remember what the first part was by the time he finished." On the other hand, "Hoang reads with expression" is a way to say Hoang seems fluent.

To increase fluency for students like Chris and Alejandro, more is the answer—more reading, that is. Practice makes, if not perfect, at least better. We have known that for a long time. But it will not help if a child is practicing the struggle, practicing the failure to coordinate word identification and comprehension. The reading practice that brings fluency has to be of the right type. To arrange the right type of practice, the teacher should be able to draw on a knowledge base from ongoing research about features of reading materials and activities relevant to fluency practice.

The well-prepared teacher knows how to find the right materials for different children and how to pass on to children the art of choosing good reading materials that provide good practice for fluency. There must be a fit between the reader and the book or passage. The fit takes into account interest and ease, familiarity and challenge, the known and what can be learned next. Is there enough but not too much demand from the vocabulary, letter-sound patterns, sentence structures? Is there help or hindrance from the use of idioms and figurative language, the reliance on inferences, the narrative or expository structure? Are there supports for chunking the text into meaningful parts—punctuation, headings, features of the typeface?

In addition, the well-prepared teacher knows how to make more difficult materials more accessible. The teacher clears hurdles likely to be encountered in a passage or chapter with minilessons on concepts, vocabulary, and text structure and reminders about patterns of letter-sound correspondence and punctuation. The well-prepared teacher arranges assisted readings, with taped or live readers, so the child follows along or echoes phrase by phrase or takes turns listening and reading with a partner. The assistant provides a model and a scaffold for fluent reading.

The teacher can arrange repeated readings so that problems with letters, words, or comprehension can be worked out gradually. Rereading can be made acceptable, even exciting, to the student practicing toward fluency. When the child reaches fluency on a passage, the child performs it—a presentation to parents, a poem for a nursing home audience, at story time for preschoolers, maybe an audio- or videotape to be made for a library of passages for use by any of the above. The performance allows for assessment of fluency, too, if the teacher sets criteria to indicate adequate speed, accuracy, and comprehension as well as an assessment of suitable emphasis and tempo.

Practice for fluency can take advantage of time beyond that allotted for reading lessons. The well-prepared teacher knows how to arrange for independent reading of materials that fit the different children and how to ensure that the practice actually happens as expected. The teacher can arrange for help from parents and the community in the process of independent, assisted, and repeated reading.

MAKING FLUENCY CENTRAL IN SECOND GRADE

A report on fluency by the Center for Improvement of Early Reading Achievement includes the following description of a second-grade curriculum that is a hothouse where children's fluency can blossom. Fourteen teachers and 105 children participated in the successful experiment. (See Stahl, Huebach, and Crammond, in press, for details.)

> The redesigned basal reading lesson used the story from the children's second-grade reading text. This text would be difficult for children reading below grade level. With the support provided by the program, however, children who entered second grade with some basic reading ability could profit from a conventional second-grade text. The teacher began by reading the story aloud to the class and discussing it. This discussion put comprehension in the foreground, so that children were aware that they were reading for meaning. Following this, the teachers reviewed key vocabulary, designed comprehension exercises, and performed other activities around the story itself. Sometimes this involved echo reading, or having the teacher read part of the story and the class or group echo it back. Other times, it involved having children read and practice part of the story. Then the story was sent home and read with the child's parents or other readers listening. For children who struggled, the story was sent home additional times during the week. Children who did not have difficulty with the story did other reading at home on these days.
>
> The next day, the children reread the story with a partner. One partner would read a page while the other would monitor the reading. Then they would switch roles until the story was finished. Following partner reading, the teacher would do some extension activities and move onto another story.
>
> Although this lesson was an important part of the program, it was not the only reading that children did. Later in the day, time was set aside for children to read books that they chose. These were usually easy to read and read for enjoyment. Children sometimes read with partners during this period, as well. This time ranged from 15 minutes in the beginning of the school year to 30 minutes by the end.
>
> Also, children were required, as part of their homework, to read at home. Outside reading was monitored through reading logs, and teachers made sure that the children in this program read at home an average of four days a week for at least 15 minutes per day.

SOURCE: M. R. Kuhn & S. A. Stahl. 2000. *Fluency: A Review of Developmental and Remedial Practices.* Ann Arbor: University of Michigan, Center for the Improvement of Early Reading Achievement.

Activity

Teacher Refresher About Phonemes

Moving Away from Spellings

This activity starts with familiar issues of spelling and sound similarities progressing to phonemes. It can bring teachers with disparate backgrounds to a common ground for subsequent work on phonemic awareness. The activity covers several sessions; there should be an instructor well versed in reading and phonology who can supply information and extra help as well as facilitate discussion and group work. Even teachers who have had little exposure to phonemes and such terminology can begin with the activities in steps 1 and 2. The discussion in step 3 may reveal the need for extra help between group meetings so that all can participate fully in the homework and the next class.

1. The whole group discusses the terms *homonym, homophone,* and *homograph.* Focus on homophones—words that are pronounced alike regardless of spelling—and homographs—words that are spelled alike regardless of how they are pronounced.

2. Divide into small groups of five or so that will also be convenient for homework groups later. For 15 minutes, small groups make lists of as many homophone and homograph pairs as they can, underlining the crucial similarity or difference in each word.

3. Reassemble in a large group. Pool the results, making a list of pairs for homophones on the blackboard or an overhead projector. Pool the results for homographs making a second list of pairs.

   Focus the whole-group discussion on the underlined parts of the words in the lists. Discuss the small units of similarity or difference in terms of phonemes. Assist the teachers to review the phonology they learned in prerequisite courses about language structure. Write a phonemic symbol for the sound that is the same in each word of the homophone pairs and the two phonemic symbols for the sounds that are different in the homograph pairs. If members of the class are familiar with different phonemic symbol systems, build consensus to rely on one notation acceptable to the group as a whole.

4. For homework each group will work on syllable frames. Consonants (C) and vowels (V) in sequence make the frames. To use a frame, vary one item, keeping the rest constant. List as many words as possible that are examples of each frame.

   - CVC—varying V, producing, for example, BIT, BEAT, for the frame where /b/ is the first constant and /t/ is the second constant;
   - CV—varying V, producing, for example, BE, BY, where /b/ is the constant;
   - C(C)VC(C)—varying C's before and after V, so where /ɛ/ is the constant, SET—also SLED and TESTS since using clusters of as many C's as possible is part of the frame.

   After discussing the frames and lists to be produced, each small group meets briefly to choose the constants to be used in each frame and to plan homework coordination. Before leaving, each group registers the constants with the instructor—no duplicates among the groups.

5. The next class meeting starts in small groups for half an hour. Each group is asked to answer the following questions, proving the answers by referring to the word lists made for homework:

   - Are there phonemes of English that can occur in only closed (CVC) syllables or only open (CV) syllables?
   - How many phonemes can there be in initial and final consonant clusters?
   - Are there any dialect differences in the class? Include evidence from words that are judged to have the same phonemes by some in the group but not everyone.

6. Reconvening as a whole group, the class pools results, checking the accuracy of the answers with respect to all the lists generated by all the groups. It may well be necessary to place conditions on generalizations due to the results from one of the frames or from dialect differences found in one of the groups. Disputes or deeper queries should be recorded and submitted to a local linguist for a response that will be distributed to the group.

A B C D E F G H I J K L M N O P Q R S T U V W X Y Z

Activity

Practicing Phonemic Awareness

This activity allows teachers to think about, experience, and create phonemic awareness activities. The activity covers multiple meetings. Articles and teachers' manuals or children's texts should be available for reference or to take home for further work. In many instances, groups that come together to learn will have different degrees and kinds of background knowledge about phonemes, so the first step is to establish common ground. This activity starts with the "interference" that good spellers have when they try to think about phonemes.

1. Background preparation in a whole-group session, recalling phonology classes:

   - Discuss words that have a different number of letters and phonemes. Write AX on the board, noting that it has two letters, three phonemes. Check to see if there is any dialect difference in the group—the most common pronunciation is likely to be /æ k s/—but there may be different pronunciations of the vowel. But the numbers of letters and phonemes are the same. Elicit more examples of words that have fewer letters than phonemes. Elicit examples of words that have fewer phonemes than letters, like MEAN (compare the different spelling for the same vowel sound in ME).

   - Write ASK on the board next to AX. Discuss pairs of words that have the same phonemes in different orders. Expect reports of dialects in which ASK and AX are homonyms, with both pronounced /æ k s/. Elicit more examples of words that have the same phonemes in different orders.

   - Establish common meaning for terms the group uses when talking about phonemes and spellings (e.g., digraph, diphthong, clusters, blends, long, short, tense, lax, nasal, affricate, open syllable).

2. Introduce phonemic awareness activities in a whole-group session.

   - Begin with phonemic manipulation. Encourage cooperation and avoid putting people on the spot. The instructor should help liberally at the start to ensure correct answers and to shift more responsibility to the class as the game progresses. Play until the class seems adept at the phoneme manipulations.

The game starts with a single-syllable spoken word (e.g., GOLD) and six player moves:

- ◆ delete the first or last phoneme (from GOLD, make OLD or GOAL)
- ❁ insert a phoneme at the beginning or the end (from OLD, make SOLD)
- ⌘ change the order of phonemes (from GOAL, make GLOW),
- ⋎ insert, delete, or change in the middle (from SOUL, make STOLE)
- ◎ pass to a new person
- 🔔 start a new chain with a new syllable

Play as a dice game with the symbols pasted to the sides of a cube that is rolled, or use a wheel to spin with segments labeled for each move. Each player rolls or spins, performs the manipulation that turns up, and passes the new syllable and a turn with the dice or wheel to the next player in the row/circle/aisle. If the dice roll is ◎, the player gets a free pass and the right to toss the syllable and the next turn to anyone in the room, even to keep it or pass it to the instructor, thereby starting a new order of players. Sometimes a sequence of deletes can end the game—no more phonemes to manipulate!

The following is a way to complicate the game to maintain interest among grown-ups:

If a player gets a syllable that is a not a word in English but just a nonsense syllable, *before* rolling the dice and following the directions on it, the player announces a choice between continuing the chain or beginning a new chain by supplying a new single-syllable word to manipulate. If the player fails to notice that the syllable is a nonsense word in English, the player loses a turn.

Discuss the concept of phonemic awareness in light of the manipulations in the game. Recall the caution about confusing phonemes and spellings. Emphasize the difference between proficiency with a language and the metalinguistics involved in phonemic awareness. Comment on any noticeable dialect differences.

3. In a whole-group session, give definitions and examples for each of the three major types of phonological awareness activities: *synthesis, analysis,* and *identity* (p.94). Provide examples of each type of activity. During the discussion, bring up the following:

   - Consider the choice of words or syllables to be used in phonemic awareness tasks. For example, prefer consonants that are least likely to be distorted if exaggerated (like /f/ or /s/ or /m/) and long vowels (as in ME) that may be easier to perceive than short vowels (as in MIT). After a good introduction to the task and some proficiency, the other phonemes might be added if phonemic awareness tasks are still needed.

   - Consider the use of manipulatives and visual aids to add a concrete element to the task. Discuss the use or introduction of letters that correspond to phonemes not only as manipulatives or visual aids but also to transfer the application of phonemic manipulations in reading and writing tasks.

   - Consider completeness of coverage. Starting with a focus on initial and final phonemes is all right, but the second elements in consonant clusters and the vowels in closed syllables shouldn't be ignored. Similarly, while two or three phoneme syllables may be a good starting point, one and four or more phoneme syllables shouldn't be overlooked.

   - Consider the needs of students. Are there dialect differences that would have an impact on the activity if certain words or phonemes are used? Is there a way to assess whether the activity is not needed at all by some children and perhaps is too difficult for others?

4. Divide the class into groups of three or four. Each group is to create a phonemic awareness activity and explain to the class the kind of unit it would be a part of. Make sure that some groups focus on synthetic, some on analytic, and some on identity activities. They may browse

through available books, articles, teachers' manuals, and children's texts for ideas. The groups will make the activity, the materials for it, and the plans to implement it. The groups will be prepared to demonstrate the activity to the class and to discuss it in terms of the matters brought up during steps 2 and 3 above.

5. Reconvene for the demonstrations and discussions. Suggest ways to pilot the activity—as improved during the discussion—with groups of children and to assess its feasibility and effectiveness.

A B C D E F G H I J K L M N O P Q R S T U V W X Y Z

Activity

Signs in a Different Alphabet

This is a group problem-solving activity that allows teachers and teachers-to-be to work with words written in an unfamiliar alphabet. The exercise can give substance to what the students have learned from reading and lectures as well as a sense of what children experience when entering alphabetic literacy.

Ask students who are literate in Russian to assist with pronunciation and to observe the problem solving.

1. The class meets first as a whole group. Explain to the students that they are going to problem solve and role play. Tell them the purpose of the activity is to have an experience somewhat similar to the complex one children have when they are learning to read. The experience will be followed by a class discussion of the multitasking involved in reading and learning to read. They have an advantage over first graders because they already know what reading is in one language. They have a disadvantage because they have not been surrounded by the Cyrillic alphabet and the Russian language before trying to read the words.

   - The problem is hunger. The solution is to identify the sign for a restaurant.

   - The role and the context: The students are tourists in a Russian city—not Moscow or St. Petersburg, a lot less cosmopolitan. Hotels, apartments, shops, restaurants, and various other establishments are in big buildings with facades that have no street-level large windows or other distinguishing marks. But there are signs.

     Show the signs the students must choose among and leave them on display.

     РЕСТОРАН БАНК ГОСТИНИЦА РЕМОНТ ТЕЛЕФОН

     The tourists didn't pay attention to signs over the past few days when they had a bilingual guide. They hadn't learned to speak, read, or write Russian before the trip. Written Russian is alphabetic, but the letters are different than the ones for English. Some look familiar from English or Greek or math.

Each tourist has learned a little Russian and found friendship groups to share problems with. In 5 minutes they'll start the groups, but first they "cram" some Russian.

Pass out small slips of paper, each one with four rows from the chart below. Mix up the rows that are given out so that no one gets the exact same four words.

Tell the students you are going to collect the papers in 5 minutes but that they can take notes to the group of friends they will work with.

Words in Russian	Hints about pronunciation	English meaning
НЕТ	nyet	not or no
НОС	nos (like 'most' without 't')	nose
НОВЫЕ	novwiy	new
ОВОШИ	ovoshi	vegetables
СЕСТРА	sestra	sister
СТАРТ	start	start
СУП	soop	soup
РЕКОРД	rekord	record
СТУЛ	stewl	chair
ЧАЙ	chie (rhymes with tie)	tea
САЛАТ	sahlaht	salad
СИГАРА	sigahrah	cigar
США	s-shah	USA
БАР	bar	bar
СЕГОДНЯ	sevodnyah	today
РАКЕТА	rocketa	rocket
ОБЕД	obed	meal

2. Divide the class into groups of five to pool resources (what they learned from the small part of the table each one had) for 30 minutes and solve the problem: Where can they get something to eat? They should talk about how they figured out what the signs said and why they think they

A B C D E F G H I J K L M N O P Q R S T U V W X Y Z

are right. Fifteen minutes into the exercise, talk to the groups about cognates—words that sound similar in two languages. Tell them that if they "sound out" the signs they might recognize the similarity to an English word that will help them out.

3. Reconvene as a whole group and give everyone a copy of the full chart above. Reveal the "answers" for each sign: БАНК (*bank*), РЕМОНТ (*repair*), ТЕЛЕФОН (*telephone*), РЕСТОРАН (*restaurant*), ГОСТИНИЦА (*hotel*). Have volunteers use their new knowledge of Cyrillic letters and Russian sounds to try pronunciations of each word. Get the "sounding out" fine-tuned by the Russian speakers.

4. Ask for volunteers to explain what answer their groups arrived at and what they did to get there. Weave in the topics in steps 5 through 8 below.

5. Discuss letter knowledge. Some letters in the Cyrillic alphabet look like ones from the Latin alphabet used for English. Consider what knowing a letter name does. Is it easier to identify, remember, refer to, and use for problem solving? Did anyone find less certainty or more errors related to letters unlike the ones in the Latin alphabet? Did anyone rely on features, orientation, and categories from the Latin alphabet to refer to letters they had no names for—ч as perhaps "upside down small h" or я as "backwards R." Notice that Б may look like a fancy lowercase b to an English reader, but Б and В are different letters in Russian. Consider youngsters figuring out what is important to notice about letters. It is a big job to remember letter shapes and distinguishing among the 26 used for English.

6. Discuss letter-sound correspondence. Elicit some specific letter-sound relations: Р is for the /r/ sound; С is for /s/; Н is for /n/. Show how the correspondence works for the Russian words in the large table. Notice the stability in letter-sound correspondences. An exception in the table is Г, but from this small set it isn't clear if there is a systematic pattern or if one of these words is an exception.

   The "tourists" know one written language and so realize that it is useful to rely on a sound-symbol correspondence from one word to read the same symbol in another word. Children have to come to that realization.

Experienced readers know, too, that it is possible to gain accuracy and speed at letter-sound correspondences and word identification. What if a child thinks it is always going to be so hard, slow, and often filled with uncertainty?

7. Discuss methods for identifying words. Was anyone aware of segmenting and blending in a left-to-right order? Did the letter-by-letter process change for any of the words? Did any bigger chunks come into play? For instance, once CT was pronounced like the beginning of the English cognate START, was it easier and faster to zoom through the rime, getting the vowel, and finding it not so hard to remember that the P in the Cyrillic alphabet is /r/ and then to recognize the final T to confirm the word identification? Did analogy from the English cognate complement the segmenting and blending process? What about the cognate "restaurant"? When a word is known through speaking and listening, the pieces seem to come together more easily and more accurately and to be more memorable. Consider the problem of a child beginning to read who does not know some supposedly easy words from their oral-language experience.

8. Discuss more general problems: Some groups will have had some missing evidence. What was pooled from their little slips of paper might have left out useful patterns shown in the larger table. Relate this to children who miss some days from school or who do not share some common childhood experiences or in some other way have gaps in their resources. Other groups will have gone down garden paths. They may have searched for related meanings to be reflected in the words—but, for instance, MEAL and VEGETABLE were not much help with RESTAURANT. How do you monitor and change directions when your approach isn't working well enough?

9. Request the class members to write about one or two things they have a new perspective on about teaching children who are learning to read. Post the responses in a class chat room for further consideration.

resources

School studies: This chapter has more detail about the teacher knowledge base than the other chapters because, while there is much publicity about the topics covered, there is insufficient information about them for effective work in teacher preparation. To counteract any idea that we believe this is the only important content, let us note here again that we call for integrated instruction with all the aspects of reading depicted on p. 6 in order to provide the opportunities for children listed on p. 11. The following references show different faces of integrated instruction:

Gambrell, L., Morrow, L. M., Neuman, S. B., & Pressley, M. 1999. *Best Practices in Literacy Instruction.* New York: Guilford.

Juel, C. 1988. Learning to read and write: A longitudinal study of 54 children from first through fourth grade. *Journal of Educational Psychology, 80,* 437-447.

Pressley, M., Wharton-McDonald, R., Allington, R., Block, C. C., Morrow, L., Tracey, D., Baker, K., Brooks, G., Cronin, J., Nelson, E., & Woo, D. 2001. A study of effective first-grade literacy instruction. *Scientific Studies of Reading, 5*(1), 35-58.

Taylor, B. M., Pearson, P. D., Clark, K. F., & Walpole, S. 1999. Effective schools/accomplished teachers. *Reading Teacher, 53*(2), 156-159.

Basic reviews of the knowledge base relevant to this chapter can be found in the following recent publications:

Kamil, M. L., Mosenthal, P. B., Pearson, P. D., & Barr, R. (Eds.). 2000. *Handbook of Reading Research: Volume III.* Mahwah, NJ: Erlbaum. (See especially Blachman on phonological awareness, Goswami on phonological and lexical processes, and Templeton and Morris on spelling.)

National Research Council. 1998. *Preventing Reading Difficulties in Young Children.* Committee on the Prevention of Reading Difficulties in Young Children, C. E. Snow, M. S. Burns, and P. Griffin, Eds. Washington, DC: National Academy Press. (See especially Part I, Chapter 2, Part II, and Part III.)

National Reading Panel. 2000. *Teaching Children to Read: An Evidence-Based Assessment of the Scientific Research Literature on Reading and Its Implications for Reading Instruction: Report of the Subgroups.* Washington, DC: National Institute of Child Health and Human Development. (See especially Chapter 2, "Alphabetics," and Chapter 3, "Fluency.")

Neuman, S. B. & Dickinson, D. K. 2001. *Handbook of Early Literacy Research.* New York: Guilford Press. (See especially Stahl on phonics and phonological awareness, Goswami on phonological development, Hiebert and Martin on beginning reading texts, Adams on explicit systematic phonics, Richgels on invented spelling, and Tabors and Snow on bilingual children.)

Language, general introductions, suitable for nonspecialists:

Finnegan, E. 1999. *Language: Its Structure and Use.* New York: Harcourt Brace.

Fromkin, V., & Rodman, R. 1998. *An Introduction to Language.* New York: Harcourt Brace.

Language, more specialized, useful background for teacher educators:

Akmajian, A., Demers, R. A., Farmer, A. K., & Harnish, R. M. 1995. *Linguistics: An Introduction to Language and Communication.* Cambridge, MA: The MIT Press.

Pinker, S. 1999. *Words and Rules: The Ingredients of Language.* New York: Perseus Books.

Language, specifically about teacher preparation:

Fillmore, L. W., & Snow, C. E. 2000. *What Teachers Need to Know About Language.* Washington, DC: Center for Applied Linguistics.

Moats, L. C. 2000. *Speech to Print: Language Essentials for Teachers.* Baltimore: Paul H. Brookes. (This textbook is specifically directed toward teacher preparation.)

Child language development:

Berko-Gleason, J., Ed. 1989. *The Development of Language.* Columbus: Merrill.

Bowerman, M., & Levinson, S. C., Eds. 2001. *Language Acquisition and Conceptual Development (Language, Culture and Cognition 3).* New York: Cambridge University Press.

Diversity within American English, overviews:

Labov, W., Ash, S., & Boberg, C. 2002. *Atlas of North American English: Phonetics, Phonology and Sound Change.* New York: Mouton de Gruyter.

Schneider, E. W., Ed. *Focus on the USA. Varieties of English Around the World, Vol. 16.* Philadelphia: John Benjamins North America.

Diversity within American English, specifically about education:

Wolfram, W., Adger, C. T., & Christian, D. 1999. *Dialects in Schools and Communities.* Mahwah, NJ: Earlbaum.

Diversity within American English, sample publications about specific varieties:

Leap, W. 1993. *American Indian English.* Salt Lake City: University of Utah Press.

Mufwene, S., Rickford, J., Bailey, G., & Baugh, J. 1998. *African-American English: Structure, History and Use.* New York: Routledge.

Phonemic awareness, phonics, and fluency, less technical and in the context of integrated instruction:

Center for the Improvement of Early Reading Achievement. 2001. *Put Reading First: The Research Building Blocks for Teaching Children to Read.* Washington, DC: National Institute for Literacy.

National Research Council. 1999. *Starting Out Right: A Guide to Promoting Children's Reading Success.* Washington, DC: National Academy Press.

Phonemic awareness, phonics, and fluency, sample publications on specific topics:

Adams, M. J., Bereiter, C., Hirshberg, J., Anderson, V., & Bernier, S. A. 1995. *Framework for Effective Teaching, Grade 1: Thinking and Learning About Print, Teacher's Guide, Part A.* Chicago: Open Court.

Adams, M. J., Foorman, B. R., Lundberg, I., & Beeler, T. 1997. *Phonemic Awareness in Young Children: A Classroom Curriculum.* Baltimore: Brookes.

Bissex, G. L. 1980. *GYNS AT WRK: A Child Learns to Read and Write.* Cambridge, MA: Harvard University Press.

Bradley, L., & Bryant, P. 1985. Rhyme and reason in reading and spelling. *International Academy for Research in Learning Disabilities, Monograph Series, 1.* Ann Arbor: University of Michigan Press.

Bruck, M., & Treiman, R. 1992. Learning to pronounce words: The limitations of analogies. *Reading Research Quarterly, 27,* 375-388.

Carlisle, J. F. 1995. Morphological awareness and early reading achievement. Pp 189-209 in L. B. Feldman (Ed.), *Morphological Aspects of Language Processing.* Hillsdale, NJ: Erlbaum.

Chomsky, C. 1979. Approaching reading through invented spelling. Pp 43-65 in L. B. Resnick and P. A. Weaver (Eds.), *Theory and Practice in Early Reading, Vol. 2.* Hillsdale, NJ: Erlbaum.

Cisero, C. A., & Royer, J. M. 1995. The development and cross-language transfer of phonological awareness. *Contemporary Educational Psychology, 28,* 275-303.

Corneau, L., Cormier, P., Grandmaison, E., & Lacroix, D. 1999. A longitudinal study of phonological processing skills in children learning to read in a second language. *Journal of Educational Psychology, 91*(1), 29-43.

Cunningham, A. E., & Stanovich, K. E. 1997. Early reading acquisition and its relation to reading experience and ability 10 years later. *Developmental Psychology, 33,* 934-945.

Dahl, K. L., Scharer, P. L., Lawson, L. L., & Grogan, P. R. 1999. Phonics instruction and student achievement in whole language first-grade classrooms. *Reading Research Quarterly, 34,* 312-341.

Ehri, L. 1992. Reconceptualizing the development of sight word reading and its relationship to recoding. Pp. 107-143 in L. E. Gough & R. Treiman (Eds.), *Reading Acquisition.* Hillsdale, NJ: Erlbaum.

Ehri, L. C. 1995. Phases of development in learning to read words by sight. *Journal of Research in Reading, 18,* 116-125.

Ehri, L. C., & Soffer, A. G. 1999. Graphophonemic awareness: Development in elementary students. *Scientific Studies of Reading, 3,* 1-30.

Ericson, L., & Juliebo, M. F. 1998. *The Phonological Awareness Handbook for Kindergarten and Primary Teachers.* Newark, DE: International Reading Association.

Fashola, O. S., Drum, P. A., & Mayer, R. E. 1996. A cognitive theory of orthographic transitioning: Predictable errors in how Spanish-speaking children spell English words. *American Educational Research Journal, 33,* 825-843.

Faulkner, H. J., & Levy, B. A. 1999. Fluent and nonfluent forms of transfer in reading: Words and their message. *Psychonomic Bulletin & Review, 6*(1), 111-116.

Foorman, B., Francis, D. J., Fletcher, J. M., Schatschneider, C., & Mehta, P. 1998. The role of instruction in learning to read: Preventing reading failure in at-risk children. *Journal of Educational Psychology, 90,* 37-55.

Gaskins, I. W., Ehri, L. C., Cress, C., O'Hara, C., & Donnelly, K. 1996/1997. Procedures for word learning: Making discoveries about words. *The Reading Teacher, 50,* 312-327.

Goswami, U. 1995. Phonological development and reading by analogy: What is analogy and what is not? *Journal of Research in Reading,* 18, 139-145.

Goswami, U. 1999. The relationship between phonological awareness and orthographic representation in different orthographies. Pp. 134-156 in M. Harris & G. Hatano, Eds., *Learning to Read and Write: A Cross-Linguistic Perspective.* Cambridge: Cambridge University Press.

Henderson, E. H. 1981. *Learning to Read and Spell: The Child's Knowledge of Words.* DeKalb: Northern Illinois University Press.

Henry, M. 1997. The decoding/spelling continuum: Integrated decoding and spelling instruction from pre-school to early secondary school. *Dyslexia, 3,* 178-189.

Hudelson, S. 1984. Kan Yu Ret an Rayt en Ingles: Children become literate in a second language. Pp. 462-477 in D. B. Durkin, (Ed.), *Language Issues: Readings for Teachers.* White Plains: Longman.

Juel, C., & Minden-Cupp, C. 1999. *Learning to Read Words: Linguistic Units and Strategies.* Ann Arbor: University of Michigan, Center for the Improvement of Early Reading Achievement.

Kuhn, M. R., & Stahl, S. A. 2000. *Fluency: A Review of Developmental and Remedial Practices.* Ann Arbor: University of Michigan, Center for the Improvement of Early Reading Achievement.

Lundberg, I., Frost, J., & Peterson, O. 1988. Effects of an extensive program for stimulating phonological awareness in preschool children. *Reading Research Quarterly, 23,* 263-284.

McKenna, M. C. 1998. Electronic texts and the transformation of beginning reading. In D. Reinking, M. C. McKenna, L. D. Labbo, & R. D. Kieffer (Eds.), *Handbook of Literacy and Technology: Transformations in a Post-Typographic World.* Mahwah, NJ: Erlbaum.

Rashotte, C., & Torgesen, J. 1985. Repeated reading and reading fluency in learning disabled children. *Reading Research Quarterly, 20,* 180-188.

Rasinski, T. V. 1990. Effects of repeated reading and listening-while-reading on reading fluency. *Journal of Educational Research, 83*(3), 147-150.

Samuels, S. J. 1979. The method of repeated readings. *The Reading Teacher, 32,* 403-408.

Share, D. 1995. Phonological recoding and self-teaching: Sine qua non of reading acquisition. *Cognition, 55,* 151-218.

Share, D., & Stanovich, K. 1995. Cognitive processes in early reading development: Accommodating individual differences into a mode of acquisition. *Issues in Education: Contributions from Educational Psychology, 1,* 1-57.

Stahl, S. A. 1992. Saying the "p" word: Nine guidelines for exemplary phonics instruction. *The Reading Teacher, 45,* 618-625.

Stahl, S. A., Duffy-Hester, A. M., & Stahl, K. A. D. 1998. Everything you wanted to know about phonics (but were afraid to ask). *Reading Research Quarterly, 35,* 338-355.

Stahl, S. A., Huebach, K., & Crammond, B. In press. Fluency-based reading instruction. *Elementary School Journal.*

Stahl, S. A., Suttles, W., & Pagnucco, J. R. 1996. The effects of traditional and process literacy instruction on first graders' reading and writing achievement and orientation toward reading. *Journal of Educational Research, 89,* 131-144.

Stanovich, K. E., & West, R. F. 1989. Exposure to print and orthographic processing. *Reading Research Quarterly, 24,* 402-433.

Tan, A., & Nicholson, T. 1997. Flash cards revisited: Training poor readers to read words faster improves their comprehension of text. *Journal of Educational Psychology, 89,* 276-288.

Torgesen, J. K., & Hecht, S. A. 1996. Preventing and remediating reading disabilities: Instructional variables that make a difference for special students. In M. F. Graves, P. van den Broek, & B. M. Taylor (Eds.), *The First R: Every Child's Right to Read.* New York: Teachers College Press.

Treiman, R. 1985. Onsets and rimes as units of spoken syllables: Evidence from children. *Journal of Experimental Child Psychology, 39,* 161-181.

Treiman, R., & Baron, J. 1983. Phonemic-analysis training helps children benefit from spelling sound rules. *Memory and Cognition, 11,* 382-389.

Vandervelden, M. C., & Siegel, L. S. 1995. Phonological recoding and phoneme awareness in early literacy: A developmental approach. *Reading Research Quarterly, 30,* 854-875.

Vandervelden, M. C., & Siegel, L. S. 1997. Teaching phonological processing skills in early literacy: A developmental approach. *Learning Disabilities Quarterly, 20,* 63-81.

Wagner, R., & Barker, T. 1994. The development of orthographic processing ability. Pp. 243-276 in V. Berninger (Ed.), *The Varieties of Orthographic Knowledge I: Theoretical and Developmental Issues.* The Netherlands: Kluwer.

Yopp, H. K. 1992. Developing phonemic awareness in young children. *The Reading Teacher, 45,* 696-703.

Yopp, H. K. 1995. A test for assessing phonemic awareness in young children. *The Reading Teacher, 49,* 20-29.

4

Moving to Success
Motivating Children to Read

For learning to read, nothing succeeds like success. Teachers see this play out in their classrooms every day. Budding readers thrill to their own mastery, cracking the complex code of letters and sounds to delve into new subjects. A child who reads well reads a lot and soon reads even better. The rich get richer.

The heartbreaker is that the child most in need of a lot of practice does not read well, does not have competence as a motivational pull toward more practice and more success. To help a child take the opportunities provided by good reading instruction, teachers often have to push-start the motivation engine for the child.

What makes a child care? What gives the zest, the energy, the drive to get to the task and get it done, to come back after an error? What is different from child to child? From task to task? What can a teacher do if a child is disinterested, distracted, or expects defeat? We may not have a single right answer, but there are ways to approach the problem, things to do, and well-prepared teachers learn them. (See the Resources at the end of this chapter for information on motivation in general and references related to reading in particular.)

grown-ups have fickle motivations

Most of all, it is important to recognize that motivations change. Everyone's motivation to read has disappeared from time to time. A terrific reader may fall asleep reading a novel, even a prize-winning one. Dust may gather on a thick, densely printed history book, although the reader started it with enthusiasm.

Motivation to *learn* to read is fickle, too, even for grown-ups. Some have never carried through on efforts to make sense of the words, symbols, maps, and charts about weather in the newspaper. For others, motivation wanes when it comes to learning to read software manuals. The fire to learn can be doused by constant comparison to more expert peers.

Even a compelling issue, like a health crisis in a family, does not guarantee persistent effort. An article on a new medical option is hard to read with its specialized vocabulary; the style of writing and the sentence structure slow the process more. The reader has to reread just about every sentence. How soon until even the most initially committed reader gives up?

Motivation lapses are seldom simply that nor are they always recognized as such. We say we were sleepy or too emotionally involved, deserved a break, needed better reading glasses, got cable TV channels for weather and history, found a computer technician, decided to rely on the doctor after all.

Whatever the case, adults have ways to conquer lapses. Boosts for flagging motivation are many and varied. Sometimes we promise ourselves a treat—a tea break after spending just 10 minutes more on the task. Other times it is a reminder of the value and the purpose of doing the reading. Sometimes what helps is our reluctance to be a quitter or confidence that any discomfort is temporary. Adults often have an assurance born of experience that, however hard it is to stay motivated, the current challenge will make them more informed and even better at reading in the long run. Even if reading and writing are not their strong points, there are other activities where adults learn that competence is the reward for trying. We have had the success that succeeds; we are the rich that get richer.

impressions of children's motivations

Recently, teachers were asked about student behaviors related to motivation: persistence, eagerness, and attention. The results indicated that many children need help with motivation.

Readers of such reports should not take it for granted that children's motivations—or aspects thereof—are static and general rather than dynamic and specific. Situations impact a child's motivations. Some children persist when reading about superheroes but not cloud formation in science. The usual lesson structure or class assignment may bring out the best in some children but not others. In an after-school setting, one child's reading effort may be enhanced by a project with peers and another might eagerly practice skills on worksheets presented by a favorite teen tutor.

We should also be aware that impressions about a child's motivation are not always to be trusted. As a thought experiment, imagine the different ways that

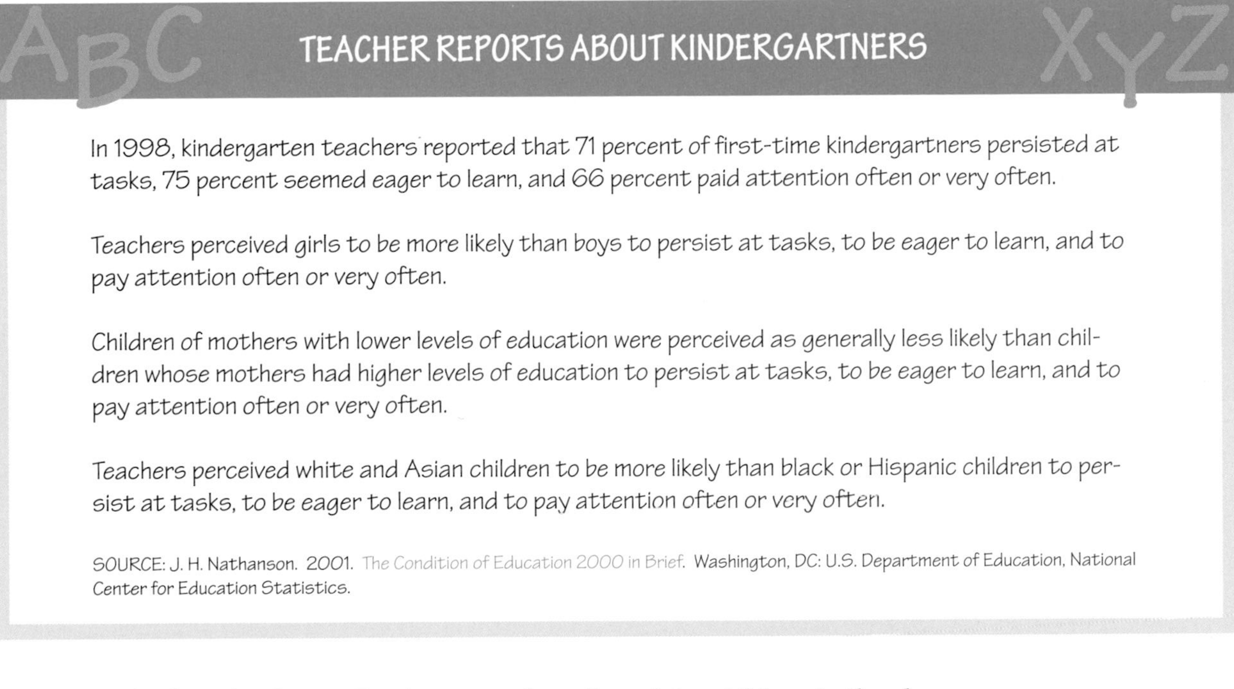

TEACHER REPORTS ABOUT KINDERGARTNERS

In 1998, kindergarten teachers reported that 71 percent of first-time kindergartners persisted at tasks, 75 percent seemed eager to learn, and 66 percent paid attention often or very often.

Teachers perceived girls to be more likely than boys to persist at tasks, to be eager to learn, and to pay attention often or very often.

Children of mothers with lower levels of education were perceived as generally less likely than children whose mothers had higher levels of education to persist at tasks, to be eager to learn, and to pay attention often or very often.

Teachers perceived white and Asian children to be more likely than black or Hispanic children to persist at tasks, to be eager to learn, and to pay attention often or very often.

SOURCE: J. H. Nathanson. 2001. *The Condition of Education 2000 in Brief*. Washington, DC: U.S. Department of Education, National Center for Education Statistics.

people might describe the motivation toward reading of the children in the class described here:

> In Boston, Massachusetts, a teacher works with a group of four- and five-year olds whose native language is Haitian Creole. The teacher is bilingual in Haitian Creole and English, having learned Haitian Creole so that she could meet the language needs of the young children in her charge.
>
> It's story time. It snowed for the first time yesterday so the teacher has found a book about a snowstorm. The children love the story. They participate as the book is being read to them. Jérémie chooses which other children he wants to play with in the snow house featured in the story. Rubenson says he wants to bring a carpet into the snow house. The children enter into the text and change it at will. The children were, in a sense, playing *in* the story.
>
> When the story features a snowman, all the children stand and gesture to show how big they would make a snowman of their own. When a picture in the book shows one child hitting another with a snowball, Rubenson tries to act it out. The children participate so fully with the book's characters that they are on the verge of taking over the story line.[1]

[1]Adapted from *Teaching Other People's Children: Literacy and Learning in a Bilingual Classroom*, by C. Ballenger (New York: Teachers College Press, 1999).

Two of Many to Motivate

The end of the first week of school in Mrs. Cobery's third grade is full of promise but serves due notice of likely challenges.

Emily struggles with reading. Her reading instruction through second grade was ineffective and unpleasant for all concerned. She reads on a first-grade level. She spent the summer avoiding books. She loves math, but the reading in word problems has started to intrude. In fact, she's on the edge of resisting everything about school.

Nina spent the summer reading every Magic Tree House book she could get her hands on. In each book, two young heroes start out reading (in a tree house, of course) but get transported into the world of the book and have fine adventures. Nina reads on grade level; these books are almost too easy for her now. The "research guide" books in the Magic Tree House series have become her favorites; they are more complex. Plenty of evidence of motivation here, but Nina can be a challenge, too. When she was supposed to work on her geography book the second day of school, she got caught devouring the latest research guide about pirates.

On Tuesday, Mrs. Cobery asked the volunteer room mother to help each child fill out a survey about attitudes, motivation, and ways to get engaged in reading. That night the teacher reviewed each child's responses.

Emily might as well have been allergic to the written word—a rash of negatives at the very mention of reading! Nina's results were more complicated. She loves reading some things but not others, and she doesn't have ways to keep herself reading when the book doesn't appeal quickly.

On Wednesday, Mrs. Cobery started journals. Each child was to write a little every day, and Mrs. Cobery would answer every afternoon in each child's journal. Only she and each student would ever read what was in his or her journal.

It's Friday now and Nina has turned the tables. She's journaling about the problems her **teacher** has with reading. Mrs. Cobery had mentioned a course she was taking at the local college: "I have to

Sitting still and paying attention? No. Motivated for the reading? Yes. Behavior that to an unprepared teacher is evidence of failing to pay attention may be the routine way that a child has learned to show eagerness at home and in the neighborhood. Following cultural rules, children spontaneously show their interest. What happens when they are trying to follow school rules? Will they know a way to express their interest? The norms children learn for behavior out of school must become part of what teachers know so that they can "see" the motivation the children display. Seeing is the first step to marshaling the children's enthusiasm toward the goal of learning to read.

Different cultures, different tasks, different topics, different children, and changes because of good teaching interventions—motivation is complex. One thing is simple, though: A child starting out needs the teacher to help with motivations and to provide motivational tools, including the fundamental springboard for success—effective instruction that makes reading the child's tool for mastering diverse subjects and chasing novel ideas.

report on a book. It is dull. Sometimes I think I will never finish it."

So Nina wrote to ask if the book is a little bit hard to understand, too. Mrs. Cobery is amused at the role switching and learns from it. Next week she'll ask Nina what is hard about third-grade geography books.

Emily's dialogue journal is quite different:

Wednesday:
"I hate this I hate this I hate this I hate this I hate this Love Emily."
"Wow, Emily, you said hate five times and then love. You surprised me. Mrs. Cobery."

Thursday:
"I hate this five times. Surprise? Love Emily."
"I am not so surprised now. Before I was surprised to see 'hate' and 'love' together. I don't expect those words together. So I was surprised. Mrs. Cobery."

Friday:
"I am so surprised now. Are you? Love Emily."

Mrs. Cobery laughed out loud. She **had** been surprised—and pleasantly—to see so clearly that Emily really looked at what her teacher wrote. Emily was following the model Mrs. Cobery gave but changing it to suit her own purposes. By Friday "hate" was gone and Emily was making a question in her own words. Mrs. Cobery knew now that she could help Emily grow in writing and find ways to teach her reading, too. So the teacher's Friday journal entry became: "I am not surprised now. What surprised you? What did you not expect? I am curious. Mrs. Cobery."

Mrs. Cobery looked back over the week in the journal and in her mind. She decided that "Love Emily" wasn't missing a comma. She didn't think of it now as a formula for closing a note. It should be an imperative, Mrs. Cobery thought, everyone should love Emily and Nina and all the rest of them.

[NOTE: Consult www.randomhouse.com/magictreehouse for more information on Magic Tree House books and research guides by Mary Pope Osborne and Weil Osborne. For various perspectives on dialogue journals, see J. Staton, R. W. Shuy, J. K. Peyton, & L. Reed, *Dialogue Journal Interaction: Classroom, Linguistic, Social, and Cognitive Views* (Norwood, NJ: Ablex, 1998).

concepts essential to motivating children

Chapters 1 through 3 call for teacher preparation with content specifically about the teaching of reading in the early grades. In contrast, the underlying issues in this chapter and the next are important for teaching any subject. Considerations specific to reading should build from the teacher's general understanding of children's motivations for school learning.

Contemporary studies related to motivation are many and varied. Teachers need to know how to evaluate claims and theories and how to make use of helpful ideas in their day-to-day practice. Teacher preparation programs should provide the time and the tools to study the basic terms and theoretical constructs related to motivation, including attitudes, interests, engagement, values, beliefs, expectations, and goals. Programs should include study of motivation in terms of specific learning tasks, perceptions of ability, and self-efficacy—the ways in which one's own choices and actions

MOTIVATION-TO-READ PROFILE

There are survey instruments to evaluate motivations or components of it. Teachers can use these to begin to investigate how a child approaches reading and learning to read. Such instruments can also help track the effectiveness of new motivational techniques teachers try in their classrooms.

Some issues can be questioned in a multiple-choice format like the following:

I tell my friends about good books I read.

- ❍ I never do this.
- ❍ I almost never do this.
- ❍ I do this some of the time.
- ❍ I do this a lot.

People who read a lot are ____________.

- ❍ very interesting
- ❍ interesting
- ❍ not very interesting
- ❍ boring

I would like for my teacher to read books out loud to the class ____________.

- ❍ every day
- ❍ almost every day
- ❍ once in a while
- ❍ never

Some topics are more open ended to be collected in individual conversations:

Tell me about your favorite author.

What are some things that get you really excited about reading books?

SOURCE: The full questionnaire from which these samples were excerpted and a study using it are reported by L. B. Gambrell, B. M. Palmer, & R. M. Coldlig, in "Assessing motivation to read with the motivation to read profile," *The Reading Teacher*, 49 (1996), 518-533.

matter. Attention should also be given to measuring and evaluating different facets of motivation and its appearance during instruction, considering, for example, the interdependence between time on task in school and a child's motivations for learning.

Motivation is not just inside a child's head or heart; it is also in the relationships among people, especially the teacher-student relationship. Teachers have to know how regulation by someone else, a teacher, contributes to children's learning and to the development of self-regulation, so that children motivate themselves well beyond lesson time. Teachers must be prepared to motivate diverse children to participate fully in varying school situations. Basic to all situations is the teacher's ability to provide a secure, positive classroom environment. At times the teacher has the leading role in the classroom scene; at other times the teacher works mostly off stage, like a producer and director, and acts only in a supporting role.

Teachers need to study motivation in conventional teacher-led lessons. Well-prepared teachers learn to choose curricula materials and to construct topics and sequences of topics that challenge but are not so difficult as to debilitate; to develop lesson structures and enact them with the pacing and student contributions that engage each child; to make adjustments for changing attention; and to use novelty, multiple approaches, and various media to attract and involve children. All this attention to what is motivating leads to learning. In turn, when a child learns, the motivational boost from the child's legitimate recognition of achievement cannot be denied. Good instruction means that the child is impressed with his or her mastery of the topic and skill involved, not focused on errors, awards, or impressing someone else.

Teacher educators should also provide opportunities for teachers to practice and understand how to organize, monitor, and enact instructional situations where the teacher is mainly in the background.

Problem-solving groups can provide a motivational base for children to practice skills and build understanding.

Another situation, play, expands attention span so that children practice skills and explore concepts painlessly. Effective teachers learn to monitor the activities and outcomes of time spent on play and peer groups, checking and changing as needed so that the expected motivations develop and the intended learning occurs. Teachers also learn to intervene from the background in a variety of ways: grouping the children carefully; enriching play and peer resources with books, field trips, or guests invited to the classroom; offering props and tools for the children to use; and

Patrick Proctor, Bilingual Resource Teacher, Waltham, Massachusetts

Making a Song and Dance of Reading and Writing

I searched for a single way to motivate students, but I didn't find a single method. My students' interests and backgrounds were pretty divergent. Something that might motivate one student wouldn't work for another.

When I was doing my master's degree, I did research in schools. I saw that the best teachers used music in their classes. They often used it as a soothing device. But when I tried it this way, it didn't work so well. Some kids found it distracting. So I looked for other ways to use music, and I eventually found that it could work well as a motivator for the children.

It all began with a song from Chile. We read the lyrics in our Spanish language arts period. The song was about a snake that terrorized a village. The lyrics were easy to read and descriptive. I had the kids respond in writing to the way the songs made them feel. Then we choreographed a sketch to go with the music. We had a whole group of kids lined up as a snake. Some kids took the parts of the villagers. We performed it in front of the whole school.

It drew on so many different things—music, costumes. There was a reading and writing component—reading the lyrics and writing about them. It spoke to a lot of things that my kids liked.

They really liked the song. It was catchy. So if the class was restless or fidgety, I'd play it, and they would automatically get quiet and focused. Every time we started working on the sketch, we began by listening to the song. They liked hearing it again and again. They read the lyrics on their own, too.

And they were particularly excited about the idea of taking something that started with lyrics and making a huge performance of it. There was a sense of anticipation. The performance went great. We did it twice.

becoming briefly involved in the scene to instruct in the use of tools and to introduce new activities, new vocabulary, and new avenues to explore.

A teacher chooses among teaching materials, lesson structures, teaching techniques, and attention-getting devices. Such choices should be informed by awareness of the potential impact on the children's later independent practice, homework, and longer-term motivations for learning.

Effective teachers master both *extrinsic* motivation (doing a task to obtain something outside the task) and *intrinsic* motivation (doing something for its own sake). They know how to use rewards as well as how to avoid misuse of them. Teachers must have an opportunity to learn what leads to a student's orientation toward mastery—where it is important to acquire skill or know more about a topic—rather than settling for situations that build an orientation toward performance—where the value is on being seen to learn well.

There is a rich web of relationships among the different aspects of the study of motivations, and well-prepared teachers have learned what is best for the long run and the short run. Teacher educators must foster teacher awareness of the possible repercussions of each motivational tactic.

About motivation as well as other matters in this book, teacher educators have to deal with groups of teachers and teachers-to-be who vary in their prior knowledge. Plans for increasing knowledge and skills dealing with motivations for teaching reading must adjust to the differences. More expert participants can be a resource for filling in gaps with those who have had less chance to learn about motivation in general.

good reading instruction raises the bar for motivating students

Teachers must be prepared to teach children to recognize when they are not understanding and to use comprehension strategies to arrive at an understanding of the text. (See Chapter 2, pp. 66-69.) Can the teacher help the child admit ignorance? Teacher educators must prepare their students for the motivational issues involved in this kind of comprehension strategy instruction. They must prepare teachers for the child who has a fear of failure and has developed a strategy of learned helplessness or one who avoids the appearance of error at all costs or one who is accustomed to being regulated by others rather than relying on self-regulation. Learning a comprehension strategy that depends on noticing and admitting confusion or errors is a major difficulty for such children. Teachers need opportunities to practice helping children develop the motivations, attitudes, and beliefs they need to make use of the metacognitive strategies that improve reading comprehension.

Throughout Chapter 3 there is a call for teachers to know about the value of children practicing outside of teacher-led lessons: practice to apply previously taught patterns of letter-sound relationships, practice to develop self-teaching of less common phonics patterns, and practice for fluent reading that connects comprehension and word identification prowess. It is the motivated child who will spend time reading. The task for the teacher is to learn enough about motivation so that each child is motivated to spend sufficient time practicing.

Teacher education programs and professional development plans should also assist teachers to examine various policies and practices in the light of knowledge about motivation. Applying the knowledge base can begin with analyzing the situations teachers and student teachers find themselves in. What are the power and the problem associated with a school mandate that each child read 25 books each year? Is there a route to self-regulation in the application of a class rule that a child read 10 pages before changing a book chosen for independent reading? These are the kinds of practical issues that can be evaluated, understood, and augmented or modified in light of studies of motivation.

the matter of materials to motivate reading

Teachers who motivate extensive and effective reading practice among students know how to provide a wide range of high-quality books and other reading materials. Quality, variety, difficulty level, and quantity all matter in a classroom library. Materials need to be interesting, engaging, and appropriately challenging.

Teachers need a chance to learn to develop and use criteria for choosing reading materials for the classroom. They cannot rely on a one-time list of titles. The stock of materials has to change to suit the grade level, the particular children, the year, and the time of year. Teachers need to know how to choose a selection that includes the classics and the contemporary, poetry and prose, fiction and nonfiction, fantasy and mystery, opinions and how-to's. A teacher must ensure that materials are available that depict a diverse range of children and families, devoid of stereotypical information. Teachers learn to include texts for the reading that is done almost without thinking as a part of everyday tasks as well as the best-quality literature that is available for youngsters. Teacher preparation sessions should help teachers look to the children, their parents, and experienced teachers to help out with hints and ideas and even to loan or donate materials.

Successful teachers learn to make suitable matches between the child and the material. The difficulty of the text—words and sentence structure—is an important consideration, but slightly harder words and more complex sentences may be accessible to a child if the text has a good layout and is accompanied by supporting figures, tables, and other graphics. The art in a book may compel a child to take on a new topic or a next level of difficulty. Teachers need an opportunity to practice with a range of materials on the same general topic, making matches with children as they improve their reading abilities.

Providing sets of books for particular topics is another skill for teachers to develop and practice. A child may have an unusual interest. Even if the teacher finds the topic unpalatable, pursuing it may be the way to engage a particularly prickly beginning reader. Projects on unusual topics are good opportunities for children to collaborate with their peers. For some children, peer relations are superior to any others as motivators. (And it lets a squeamish teacher off the hook a bit.) Teachers need not be experts in all areas, but they should know how to support interests outside their own expertise. They should be able to suggest a wide range of topics and to provide books, computer programs, web links, visitors, and field trips—whatever works to start and then maintain interest. Effective teachers know how to work with children's peer groups so that the focus and quality of the learning meet high expectations.

Through collaboration with more expert teachers, novices may well learn that the best reading material to bring to the classroom at some point in time might be something with no words or letters, like an ant farm. If the topic that attracts is ants, bring in the ants. If the topic is trains, find room for a toy train set in a corner of the classroom. Material to motivate reading isn't always printed matter. That can be added when it is time to figure out what the ants need to thrive or what stations the Grand Crescent Express has to stop at.

can the routine aspects of reading be interesting?

Good teacher preparation and the use of good materials allow the teacher to pace lessons and sequence tasks so that children make progress and avoid being discouraged. Even so, many children express their discontent with lessons on the mechanics of reading, and many teachers cannot imagine getting into interesting conversations with children about spellings and decoding.

When it comes to a story, many teachers feel confident, even inspired, to entice children to participate. They stir curiosity about the goals of a character or the intricacy of a plot. They evoke feelings related to a setting, theme, or mood. Their conversations with children about stories go far to motivate children to finish reading a particular story started with the teacher and, in the long run, to read more and more on their own.

But these same teachers may have few tricks of the trade when it comes to providing motivational boosts for children bogged down in the details of the written language and the mechanics of learning to read. Skilled teachers have had a chance to learn and practice how to do just this, exploiting children's creativity and curiosity in explorations of the writing system. They know how to create lessons that are neutral grounds—avoiding anxiety from past history of success or failure—where teachers can converse about letters and sounds and create opportunities to change attitudes and motivations.

writing motivates reading

A teacher who learns to capture a child's talk in writing has learned another basic way to provide motivation for reading and learning to read. A child can find out that what is said lasts when it is written down, that when reading it a friend or even a stranger not only can but *has to* say whatever the child wrote (or dictated for the teacher to write). The writing side of literacy brings the realization of its power, and a child awakening to that power has motives for writing and reading.

Well-prepared teachers know how to model writing and assist beginners to write on their own, making sure to take the time to read again what has been written. The successful teacher knows about providing writing materials in play centers—to

My Grandpa and Dad Going Fishing

Here is Grandpa and me carrying our fishing poles. My Grandpa, my Dad and I went to the lake. The lake was called Wright's Lake. We went out on a boat. It was a canoe.

We went fishing. I fish with a fishing pole. We used worms for bait. We put the worms on the fishing hook. Then we throw the hook in. Then we wind it up and wait until the fish come. While we wait we talk quietly. If we talk real loud the fish would swim away.

We had to wait for a long time. We didn't catch any fish.

write orders in a restaurant, take attendance in a school, make signs for the roads in the block corner or sandbox (see also Chapter 1). In teacher-led sessions, an expert teacher knows how to tie practice with letter names and shapes to making labels for people and things in the room; on another day she has the group remember a recent field trip and then compose sentences for a group essay on their experience. One on one, she prints the title under a drawing a child describes and writes down the story that a child makes up.

For older children, teachers have to learn to provide multiple purposes for writing. A curriculum unit in any subject may culminate in writing and illustrating books that the children then read aloud to kindergarten buddies. After teaching about the history and geography of the town, the teacher might arrange for the students to write a web page and answer e-mail sent in about it. Techniques like these require effective teacher study sessions.

Teachers should have a chance to learn about ways to embed writing in the daily routines of the classroom, too. It takes planning and practice, but teachers can learn how to introduce and manage journals, especially dialogue journals in which the child and the teacher share space for extended written conversations, some of which can be about why it is good to practice reading!

Finally, teachers need opportunities to learn more about tools for writing. It may be well known that there are decisions to make about how to provide independent and orderly access to white boards and markers, various types and sizes of paper, pencils, rubber stamps, staplers, tape, etc. Today's teachers, however, also need to learn to use a range of computer supports for writing and publishing—idea organizers and story starters, for example, and aids for spelling and word choice or for adding illustrations before printing and publishing a story or an essay.

teaching directly about motivation to read

There are interesting children's books that teachers can learn to use as cornerstones for lessons specifically about why someone puts time and effort into learning to read. Two are by Maria Gordon—*Dogs Can't Read: Why We Learn to Read* and *Mice Can't Spell: Why We Learn to Write* (Littleton, MA: Sundance, 2000). Another example is *The Old Woman Who Loved to Read* by John Winch (New York: Holiday House, 1997). Teachers must learn to draw attention to the characters in such books and to use the problems and resolutions in the plot as a way to help children recognize their own motivational challenges and victories.

Teachers can learn to use questionnaires to plan and conduct even more direct lessons about reading motivation. Children can be asked about their habits, opinions, and preferences. One questionnaire uses Garfield, the cartoon cat, to probe

5. How do you feel about spending free time reading?

6. How do you feel about starting a new book?

7. How do you feel about reading during summer vacation?

8. How do you feel about reading instead of playing?

attitudes about reading (McKenna & Kear, 1990). The topics include how the child feels about books as presents, about the approach of reading time, and so forth. In professional development sessions, teachers can learn to exploit such instruments as the basis for lessons that teach children to think about why they do what they do and how to change if they need to.

enlisting families to motivate readers

Home can be the bedrock for reading motivation. Some parents initiate contact with schools, filled with questions and ideas. But not all families come to the teacher easily and comfortably. Teachers must have knowledge and strategies to work with families of all different kinds.

A teacher can tip the scale to increase family involvement in a child's reading development. Increased family involvement yields increased student achievement. Despite the power of these two facts taken together, too few teachers are prepared well enough to enlist families in children's reading development.

A teacher who learns to involve families has allies. The family provides extra eyes, ears, and problem-solving resources to figure out the motivational puzzles that children present at different times and about different tasks throughout the school year.

culture and motivation

Reading practices vary from culture to culture—what to read, when to read, how to show appreciation and interest in reading. Many children today experience a great divide between the culture at home and the dominant culture at school. Culture has been described as a set of survival rules—what to do to succeed as a person. It is a cultural matter whether expertise at reading and writing carries much weight in the making of a good person. Reading might be seen as central or peripheral, optional or required.

Schools cannot mimic or mirror all the cultures represented in their student bodies. A teacher need not match each child's culture, but the teacher does need to know about it. Teachers certainly need to learn how culture comes to bear on behaviors within reading lessons.

Effective teachers figure out if something unexpected from their point of view is a quirk, a passing fancy, or a result of a custom or value with deep roots in a student's cultural identity. Recognizing a cultural pattern is only the beginning. Teacher education and professional development must also foster a teacher's ability to choose an effective teaching strategy—be it to use the cultural response, adapt it, avoid it, or find a way to coexist with it.

Well-prepared teachers do not leave it to the young child to work out cultural conflicts on his or her own. The child's culture provides an image of a successful person and guidance for a child to become one; the teacher and the school must ensure that there is room for reading in that image and that the child attains success.

How can teacher educators increase teacher knowledge and skills about cultural matters outside their own personal experience? There are published case studies in which cultural expectations come into play in the teaching of reading. Some are on video or in multimedia formats. The key for effective teacher preparation is to promote discussion and reflection on the cases, to tie them to specific methods or aspects of teaching reading, and to examine how particular lesson plans may be affected.

Overall, children's motivation to read and to learn to read can take many guises. It can wax and wane. A teacher with cultural sensitivity and assistance from families can have an impact on it. Recognizing that competence is the best motivator, teachers work hard to increase each child's reading achievement. As a part of that work, the teacher knows how to push-start the motivation engine so that the child's independent competence can turn over and kick in.

Activity

What and Why Do You Read?

Reading materials made available in the classroom should go beyond books. This activity keys teachers into the different situations that motivate reading. It can improve their ability to stock the classroom with diverse materials for reading so that different children can become interested.

The activity begins with "homework"—a specialized diary for a week. Discussion groups analyze the diaries and report results to the whole class to discuss motivations for reading. In another session, small groups brainstorm about materials for classroom use.

The instructor or facilitator asks teachers or teachers-to-be to keep a reading diary for one week, using the following rules:

- Days 1 and 2: Just before bedtime, write down what you read and how long you spent reading.

- Days 3 and 4: Set an alarm to interrupt your ordinary life every 2 hours to make a note about what you read and how long you spent reading.

- Days 5 and 6: Pick a couple of 1-hour periods to document carefully, interrupting to note every 10 minutes if there was or was not any reading going on.

Divide the class into small groups to meet for an hour or two. The instructor or facilitator should circulate to stimulate the discussions.

- Day 7: Pool diaries and prepare a report to the class. List the motivational triggers that got the diarists to read. Expect that there will be short mundane bursts of reading—getting an 800 number from a TV commercial or consulting a directory to find where in the building the new dentist's office is. There will also be longer times spent on informative and entertaining reading. Most people will have ignored some of the mundane reading until they interrupted their days to make their diary entries.

About a week later, convene the class as a whole to share group results. The class will find great diversity in the motivational triggers for reading, not only among class members but also within the experiences of each person. Consider which triggers have been seen to occur in classrooms and which have not had a chance to appear but might be helpful for work with some students. Discuss how increasing the variety of materials and activities can be a resource for solving the variety of motivational puzzles each class of children presents to a teacher.

The next week reconvene as small groups. Problem solve about where to get materials and how to set situations and tasks so that children will use them. On a class bulletin board or in a teachers' lounge, post ideas and sources of materials for other teachers to read and use. Update postings as the approaches are used and report on evaluations of effectiveness.

Activity

Exploring spellings

In this activity, teachers study lessons that get children to collect and observe data, analyze patterns, and report on interesting conclusions about letter-sound patterns in English. Occasional lessons like these do *not* substitute for systematic curricula for decoding words and spelling. They are intended to allow children and teachers to step back, see the interesting parts of the writing system, and return with renewed appetite to the regular curriculum that will work toward automaticity in word recognition.

Divide the class of teachers into two groups. Each group is asked to develop a lesson and the materials for it and to role play it. Give each group a general description of a lesson:

Lesson 1, to be used with second or third graders in small groups while the teacher roams: C corresponds to different sounds in words like *cougar* and *ceiling*. Is it random or is there a pattern? Give students dictionaries to use. If needed, suggest they make separate lists for cases where the C sounds like /s/ and where it sounds like /k/. Prompt students to think about C's that are not the first letters of words. If needed, help them notice that the dictionary is less useful for gathering data about word middles. Help students find another data source, perhaps passages from textbooks or magazines, and add words to the lists. When students have about 30 words on each list, prompt them to sort the lists for patterns. If needed, ask what never or hardly ever happens in one list but happens a lot in another list. If needed, draw attention to the different vowels that follow the C. Help them to form hypotheses, test them on new data, adjust the hypotheses as needed, and publish the methods and conclusions on bulletin boards in the classroom or on the World Wide Web.

Lesson 2 to be used with kindergarten and first graders in a teacher-led group: S and Z have quite a bit in common and so do the sounds most often associated with them (/s/ and /z/). But they are different, too. Lead a discussion about similarities between the letters. Elicit differences between the letters. Help the children to notice the direction of movement across the page and to focus on the points on Z that contrast with the curves in S. Summarize the comparison and contrast. Have children say words beginning with the two sounds /s/ and /z/ and then isolate only the first sounds. Focus on what is the same about the sounds. (Help children to think of hissing and to notice where the front part of the tongue is.) Then switch to the differences. Instruct children to put fingertips lightly on their throats when saying each sound. Help children notice the throbbing that happens with /z/ but not with /s/. Summarize the comparison and contrast.

While the groups of teachers or teachers-to-be are working on their plans, materials, and role play, the instructor should coach with suggestions about materials and discussion strategies and foresight about possible pitfalls. There are routines for getting attention and working well in cooperative groups that should be considered. And, of course, there are concerns for the "data": Can all the children see the posters or pictures that are used? If they are supposed to focus on a small part of the word, how can it be made prominent enough for them? For children who will not or cannot read the words in the data, how can the lesson tasks be structured so that they can contribute?

Especially important is the discovery conversation in the lessons. It takes practice and being coached to encourage exploration but avoid pandemonium. Ideas should definitely come from the children's observations, but there is a "time to tell" when the discussion is getting bogged down.

Have the groups revise the lessons and materials as needed after the role play. Ask them to prepare a report that includes (1) the lesson plan, (2) directions for materials used, (3) hints about pitfalls and how to avoid them, and (4) ideas about other topics that could follow the same general lesson plan (the sounds of words with G in them; the letters U and W.)

Post the successful reports to be shared with other teachers and teachers-to-be.

A B C D E F G H I J K L M N O P Q R S T U V W X Y Z

Activity

The Family Connection

Teachers learn by coordinating with other teachers and using outside resources in order to involve families in their effort to improve reading achievement. This school-based activity works out a long-term plan, evaluates the intermediate effects, designs new tactics, and maintains the effort over time.

1. A month before the first formal sessions a facilitator develops a questionnaire or a structured interview to collect information on experiences that teachers in the school have had with family involvement. The results are analyzed and reported on—goals and outcomes, success and problems, materials and measures of effectiveness. The facilitator sets up a resource area for the family connection project so that teachers can drop by and consult reports and materials.

2. Teams of teachers take about a month to prepare reports for the school community based on a published resource. The team reports are added to the resource center. The facilitator helps the teams search the Internet and libraries for articles and books as well as assemble a library of resources to start with, for example:

   Epstein, J. L. 1991. Effects on student achievement of teachers' practices of parent involvement. *Advances in Reading/Language Research, 5*, 261-276.

   Epstein, J. L. 1995. School/family/community partnerships: Caring for the children we share. *Phi Delta Kappan, 76*(5):701-712.

   Funkhouser, J. E., & Gonzales, M. R. 1997. *Family Involvement in Children's Education: Successful Local Approaches: An Idea Book.* Washington, DC: National Institute on the Education of At-Risk Students, Office of Educational Research and Improvement, U.S. Department of Education.

   Shartrand, A. M., Weiss, H. B., Kreider, H. M., & Lopez, M. E. 1997. *New Skills for New Schools: Preparing Teachers in Family Involvement.* Cambridge, MA: Harvard Family Research Project, Harvard Graduate School of Education.

   Partnership for Family Involvement in Education publications, including *Partners for Learning: Preparing Teachers to Involve Families and Building Support for Better Schools: Seven Steps to Engaging Hard-to-Reach Communities* (*www.pfie.ed.gov*).

3. The facilitator contacts local Head Start and Even Start or other family literacy programs to learn about and build on already developed family

connections for literacy education. Contact information and materials are added to the resource center.

4. The facilitator collects, for the resource center, samples of materials that can be sent to homes as part of family involvement efforts. One example is *School-Home Links Reading Kits,* with activities designed to help families reinforce the reading and language arts skills their children are learning at school. There are books for different grade levels, starting with kindergarten, in English and Spanish. Available at *www.pfi.ed.gov,* they are a joint project of the U.S. Department of Education, the Corporation for National Service, the *Los Angeles Times,* and Little Planet Learning. There is also a series of videotapes for family use called *Take Me to Your Readers.* There are four tapes for parents of children in kindergarten on up: *Reading About Books; Talking About Books; Building Vocabulary and Knowledge; Cracking the Code with Games and Writing.* For more information on the tapes, contact Little Planet Learning (telephone: 888-974-2248).

5. At a first meeting of all the teachers, the facilitator describes the success so far—the reports and other materials available in the new resource area for family connections. The teams of teachers are asked to study the materials and to develop and make a modest effort for 2 months and then report back to the group, so that a schoolwide plan can be developed. They are asked to take care to describe how they know if their plan worked well and to document problems or changes that came up along the way and new resources they found or invented.

6. After a few months, brief written reports are made available in the resource center, and a second large meeting is called to discuss the reports and plan a longer-term school effort. The essential message is that problems encountered are opportunities to develop new approaches and to refine plans. This time plans are made for 5-month trials, and the process is repeated. Semiannual reviews of parental involvement efforts can become a part of the school culture and, with regular additions from teacher study teams, the resource center for family involvement will be a well-nurtured part of the institution.

Activity

Videotape with a Cultural Lens

Literally using a second impression, teachers-to-be become aware of how much first impressions are shaped by cultural expectations. Guided discussion of a video leads teachers to see what it takes to recognize, establish, or maintain reading motivation with children from different cultural backgrounds.

Choose a video case. Have a third party view it to be sure that it clearly illustrates the points intended for discussion. Try to enlist the help of a knowledgeable member of the culture depicted, to choose and edit the video segment, and, if possible, to participate in the discussion. The video should be brief, about 10 minutes.

Prepare a viewer's guide to the videotape. It should have 5 to 10 short statements about the case with a space for marking true or false. Add one open-ended prompt for a short note about one of the children depicted. The questions should be asked twice—to be answered during a first viewing and then a second. The following guide would work for a video of the lesson with Haitian Creole children described briefly on p. 121:

Viewing Guide: Mark T (for true) or F (for false)

While watching the first time:

____ Teacher uses children's home language.

____ Students are impulsive.

____ Teacher takes account of cultural needs of the children.

____ Students appear motivated to listen to the story.

____ Teacher needs more control of the class before trying to read a story aloud.

____ Teacher uses a book that is relevant to the children.

____ Students are really engaged in the activity.

____ Teacher has effectively prepared an activity for this group of children.

Briefly describe Jeremie's or Rubenson's engagement in the book-reading activity.

While watching the second time:

____ Teacher uses children's home language.

____ Students are impulsive.

____ Teacher takes account of cultural needs of the children.

____ Students appear motivated to listen to the story.

____ Teacher needs more control of the class before trying to read a story aloud.

____ Teacher uses a book that is relevant to the children.

____ Students are really engaged in the activity.

____ Teacher has effectively prepared an activity for this group of children.

Briefly describe the same child's engagement in the book-reading activity.

Introduce the case by providing general information about the lesson, the teacher, and the children.

Give each person a copy of a viewer's guide. Show the tape one time, asking them to fill in the blanks on the left side of the guide.

Discuss the questions on the guide and the responses. The instructor (and a visitor from the culture if possible) gradually brings in relevant ethnographic information about the culture of the children in the video.

Show the videotape a second time with the students completing the second half of the handout.

Briefly discuss ways in which their responses changed from the first viewing to the second viewing. Encourage the teachers to apply this experience to their own lives. They can share information about reading lessons they have witnessed, conducted, or participated in where cultural differences played an acknowledged or unacknowledged role. They can make plans about actions in their own classrooms in the future.

resources

For our purposes it is sufficient to use the term *motivation* loosely, covering affective, cognitive, social, and cultural matters with many components and variations. There is much value in more precise treatments for teacher preparation, and below we provide additional resources that bring more depth and breadth to the topics in Chapter 4:

General issues about motivation and learning, book-length treatments:

McInerny, D., & S. Vanetten (Eds.). 2001. *Research on Sociocultural Influences on Motivation and Learning.* Greenwich, CT: Information Age Publishing.

Meichenbaum, D., & Biemiller, A. 1998. *Nurturing Independent Learners: Helping Students Take Charge of Their Learning.* Cambridge, MA: Brookline.

Pintrich, P., & M. Maehr (Eds.). 1995. *Advances in Motivation and Achievement, Vol. 9.* Greenwich, CT: JAI Press.

Stipek, D. J. 2002. *Motivation to Learn: Integrating Theory and Practice, 4th Edition.* Boston: Allyn & Bacon.

General issues about motivation and learning, articles:

Cameron, J., & Pierce, W. D. 1994. Reinforcement, reward, and intrinsic motivation: A meta-analysis. *Review of Educational Research, 64,* 363-423.

Deci, E. L., Koestner, R., & Ryan, R. M. 1999. A meta-analytic review of experiments examining the effects of extrinsic rewards on intrinsic motivation. *Psychological Bulletin, 125,* 627-668.

Eccles, J. S., Wigfield, A., & Schiefele, U. 1998. Motivation to succeed. Pp. 1017-1098 in W. Damon (Editor-in-Chief) and N. Eisenberg (Vol. Ed.), *Handbook of Child Psychology: Vol. 3. Social, Emotional, and Personality Development* (5th ed.). New York: Wiley.

Pressley, M. 1995. More about the development of self-regulation: Complex, long-term, and thoroughly social. *Educational Psychologist, 30,* 207-212.

Rigby, C. S., Deci, E. L., Patrick, B. C., & Ryan, R. M. 1992. Beyond the intrinsic-extrinsic dichotomy: Self-determination in motivation and learning. *Motivation and Emotion, 16,* 165-185.

Stipek, D., Feller, R., Daniels, D., & Milburn, S. 1995. Effects of different instructional approaches on young children's achievement and motivation. *Child Development, 66,* 209-221.

Reading and motivation issues, book-length treatments:

Baker, L., Dreher, M. J., & Guthrie, J. T. (Eds.). 2000. *Engaging Young Readers: Promoting Achievement and Motivation.* New York: Guilford Press.

Cunningham, P. M., & Allington, R. L. 1999. *Classrooms that Work: They Can All Read and Write, 2nd ed.* New York: Harper Collins.

Gambrell, L., & Almasi, J. F. 1996. *Lively Discussions! Fostering Engaged Reading.* Newark, DE: International Reading Association.

Gambrell, L., Morrow, L. M., Neuman, S. B., & Pressley, M. 1999. *Best Practices in Literacy Instruction.* New York: Guilford.

Guthrie, J. T., & Alvermann, D. (Eds.). 1999. *Engagement in Reading: Processes, Practices, and Policy Implications.* New York: Teachers College Press.

Guthrie, J., & A. Wigfield (Eds.). 1997. *Reading Engagement: Motivating Readers Through Integrated Instruction.* Newark, DE: International Reading Association.

Nell, V. 1988. *Lost in a Book: The Psychology of Reading for Pleasure.* New Haven, CT: Yale University Press.

Verhoeven, L., & Snow, C. (Eds.). 2001. *Literacy and Motivation: Reading Engagement in Individuals and Groups.* Mahwah, NJ: Erlbaum.

Reading and motivation issues, articles:

Alexander, P. A., & Murphy, P. K. 1998. Profiling the differences in students' knowledge, interest, and strategic processing. *Journal of Educational Psychology, 90*(3), 435-447.

Baker, L., & Wigfield, A. 1999. Dimensions of children's motivation for reading and their relations to reading activity and reading achievement. *Reading Research Quarterly, 34,* 452-477.

Cohen, S. G., McDonell, G., & Osborn, B. 1989. Self-perceptions of "at-risk" and high achieving readers: Beyond Reading Recovery achievement data. Pp. 117-122 in S. McCormick & J. Zutell (Eds.), *Cognitive and Social Perspectives for Literacy Research and Instruction.* Chicago: National Reading Conference.

Dixon-Krauss, L. A. 1995. Partner reading and writing: Peer social dialogue and the zone of proximal development. *Journal of Reading Behavior, 27*(1), 45-63.

Fisher, C. W., Berliner, D. C., Filby, N. N., Marliave, R., Cahen, L. S., & Dishaw, M. M. 1980. Teaching behaviors, academic learning time and student achievement: An overview. Pp. 7-32 in C. Denham & A. Lieberman (Eds.), *Time to Learn.* Washington, DC: National Institute of Education.

Fresch, S. L. 1995. Self-selection of early literacy learners. *Reading Teacher, 49,* 220-227.

Guthrie, J. T. 1996. Educational contexts for engagement in literacy. *Reading Teacher, 49,* 432-445.

Hepler, S. 1998. Nonfiction books for children: New directions, new challenges. Pp. 3-17 in R. A. Bamford & J. V. Kristo (Eds.), *Making Facts Come Alive: Choosing Quality Nonfiction Literature K-8.* Norwood, MA: Christopher-Gordon.

Martinez, M., Roser, N. L, Hoffman, J. V., & Battle, J. 1992. Fostering better book discussions through response logs and a response framework: A case description. Pp. 303-311 in C. K. Kinzer & D. J. Leu (Eds.), *Literacy Research, Theory, and Practice: Views from Many Perspectives.* Forty-first Yearbook of the National Reading Conference. Chicago: National Reading Conference.

Mason, J. M., Peterman, C. L., Powell, B. M., & Kerr, B. M. 1989. Reading and writing attempts by kindergartners after book reading by teachers. Pp. 105-120 in J. M. Mason (Ed..), *Reading and Writing Connections.* Boston: Allyn & Bacon.

McKenna, M. C., & Kear, D. J. 1990. Measuring attitude toward reading: A new tool for teachers. *The Reading Teacher, 43,* 626-639.

McKenna, M. C., Kear, D. J., & Ellsworth, R. A. 1995. Children's attitudes toward reading: A national survey. *Reading Research Quarterly, 30,* 934-956.

Meece, J. L., & Miller, S. D. 1999. Changes in elementary school children's achievement goals for reading and writing: Results of a longitudinal and an intervention study. *Scientific Studies of Reading, 3*(3), 207-229.

Metsala, J. L., Wigfield, A., & McCann, A. D. 1996. Children's motivations for reading. *The Reading Teacher, 50,* 360-362.

Oldfather, P., & Dahl, K. 1994. Toward a social constructivist reconceptualization of intrinsic motivation for literacy learning. *Journal of Reading Behavior, 26*(2), 139-158.

Reeve, J., Bolt, E., & Cai, Y. 1999. Autonomy-supportive teachers: How they teach and motivate students. *Journal of Educational Psychology, 91*(3), 537-548.

Schiefele, U. 1999. Interest and learning from text. *Scientific Studies of Reading, 3*(3), 257-279.

Skinner, E. A., Wellborn, J. G., & Connell, J. P. 1990. What it takes to do well in school and whether I've got it: A process model of perceived control and children's engagement and achievement in school. *Journal of Educational Psychology, 82*(1), 22-32.

Stahl, S. A., Suttles, W., & Pagnucco, J. R. 1996. The effects of traditional and process literacy instruction on first graders' reading and writing achievement and orientation toward reading. *Journal of Educational Research, 89,* 131-144.

Sweet, A. P., Guthrie, J. T., & Ng, M. M. 1998. Teacher perceptions and student reading motivation. *Journal of Educational Psychology, 90*(2), 210-223.

Turner, J. C. 1995. The influence of classroom contexts on young children's motivation for literacy. *Reading Research Quarterly, 30,* 410-441.

Wigfield, A., & Guthrie, J. T. 1997. Relations of children's motivation for reading to the amount and breadth of their reading. *Journal of Educational Psychology, 89,* 420-432.

Books related to families, diversity of culture, and community:

Au, K. 1993. *Literacy Instruction in Multicultural Settings.* Orlando, FL: Harcourt Brace.

Baker, L., Afflerbach, P., & Reinking, D. (Eds.). 1996. *Developing Engaged Readers in School and Home Communities.* Mahwah, NJ: Erlbaum.

Gregory, E. (Ed.). 1997. *One Child, Many Worlds: Early Learning in Multicultural Communities.* Language and Literacy Series. Williston, VT: Teachers College Press.

Taylor, D., & Dorsey-Gaines, C. 1988. *Growing Up Literate: Learning from Inner-City Families.* Portsmouth, NH: Heinemann.

Taylor, D., and Strickland, D. 1986. *Family Storybook Reading.* Portsmouth, NH: Heinemann.

Articles related to families, diversity of culture, and community:

Au, K. 1997. Ownership, literacy achievement, and students of diverse cultural backgrounds. Pp. 168-182 in J. T. Guthrie & A. Wigfield (Eds.), *Reading Engagement: Motivating Readers Through Integrated Instruction.* Newark, DE: International Reading Association.

Duke, N. K. 2000. For the rich it's richer: Print environments and experiences offered to first-grade students in very low- and very high-SES school districts. *American Educational Research Journal, 37,* 456-457.

Edwards, P. A. 1995. Connecting African-American families and youth to the school's reading program: Its meaning for school and community literacy. Pp. 263-281 in V. L.

Gadsden & D. Wagner (Eds.), *Literacy Among African-American Youth: Issues in Learning, Teaching, and Schooling.* Cresskill, NJ: Hampton Press.

Elliot, J. A., & Hewison, J. 1994. Comprehension and interest in home reading. *British Journal of Educational Psychology, 64,* 203-220.

García, E. E. 1994. Linguistically and culturally diverse children: Effective instructional practices and related policy issues. Pp. 65-86 in H. C. Waxman, J. Walker de Félix, J. E. Anerson, & H. P. Baptiste (Eds.), *Students at Risk in At-Risk Schools: Improving Environments for Learning.* Newbury Park, CA: Corwin Press.

Goldenberg, C. N., & Gallimore, R. 1995. Immigrant Latino parents' values and beliefs about their children's education: Continuities and discontinuities across cultures and generations. Pp. 183-228 in P. Pintrich & M. Maehr (Eds.), *Advances in Motivation and Achievement, Vol. 9.* Greenwich, CT: JAI Press.

Goldenberg, C., Reese, L., & Gallimore, R. 1992. Effects of school literacy materials on Latino children's home experiences and early reading achievement. *American Journal of Education, 100,* 497-536.

Madrigal, P., Cubillas, C., Yaden, D. B., Jr., Tam, A., & Brassell, D. 1999. *Creating a Book Loan Program for Inner-City Latino Families.* Ann Arbor, MI: Center for Improvement of Early Reading Achievement.

Neuman, S. B., & Celano, D. 2001. Access to print in low-income and middle-income communities. *Reading Research Quarterly, 36*(1), 8-27.

Neuman, S. B., & Roskos, K. 1992. Literacy objects as cultural tools: Effects on children's literacy behaviors in play. *Reading Research Quarterly, 27,* 202-225.

Neuman, S. B., & Roskos, K. 1993. Access to print for children of poverty: Differential effects of adult mediation and literacy-enriched play settings on environmental and functional print tasks. *American Educational Research Journal, 30,* 95-122.

Purcell-Gates, V. 1996. Stories, coupons, and the TV guide: Relationships between home literacy experiences and emergent literacy knowledge. *Reading Research Quarterly, 31,* 406-428.

Rueda, R., MacGillivray, L., Monzo, L., & Arzubiaga, A. 2001. Engaged reading: A multilevel approach to considering sociocultural factors with diverse learners. Pp. 233-264 in D. McInerny & S. Vanetten (Eds.), *Research on Sociocultural Influences on Motivation and Learning.* Greenwich, CT: Information Age Publishing.

Schmidt, P. 1995. Working and playing with others: Cultural conflict in a kindergarten literacy program. *The Reading Teacher, 48,* 404-412.

Strickland, D. S., & Taylor, D. 1989. Family storybook reading: Implications for children, families, and curriculum. Pp. 27-34 in D. S. Strickland and L. M. Morrow (Eds.), *Emerging Literacy: Young Children Learn to Read and Write.* Newark, DE: International Reading Association.

Taylor, D. 1986. Creating family story: "Matthew! We're going to have a ride!" Pp. 139-155 in W. H. Teale & E. Sulzby (Eds.), *Emergent Literacy: Writing and Reading.* Norwood, NJ: Ablex.

Thompson, R., Mixon, G., & Serpell, R. 1996. Engaging minority students in reading: Focus on the urban learner. Pp. 43-63 in L. Baker, P. Afflerbach, & D. Reinking (Eds.), *Developing Engaged Readers in School and Home Communities.* Mahwah, NJ: Elbaum.

Volk, D. 1997. Questions in lessons: Activity settings in the homes and school of two Puerto Rican kindergartners. *Anthropology & Education Quarterly, 28*(1), 22-49.

Using cases in teacher preparation:

Alexander, N. P. 2000. *Early Childhood Workshops that Work: The Essential Guide to Successful Training and Workshops.* Beltsville, MD: Gryphon House.

Bransford, J., Kinzer, C., Risko, V., Rowe, D., & Vye, N. 1989. Designing invitations to thinking: Some initial thoughts. *National Reading Conference Yearbook, 38,* 35-54.

Hiebert, J., Gallimore, R., & Stigler, J. W. 2002. A knowledge base for the teaching profession: What would it look like and how can we get one? *Education Researcher, 31,* 5, 3-15.

Hughes, J. E., Packard, B. W., & Pearson, P. D. 1997. Reading classroom explorer: Visiting classrooms via hypermedia. Pp. 494-506 in C. K. Kinzer, K. A. Hinchman, & D. J. Leu (Eds.), *Inquiries in Literacy Theory and Practice.* Chicago: National Reading Conference.

Hughes, J. E., Packard, B. W., & Pearson, P. D. 2000. Pre-service teachers' experiences using hypermedia and video to learn about literacy instruction. Available online at *www.ciera.org/library/archive/2000-06/art-online-00-06.html.*

Kinzer, C. K., & Risko, V. J. 1998. Multimedia and enhanced learning: Transforming preservice education. Pp. 185-202 in D. Reinking, M. C. McKenna, L. D. Labbo, & R. D. Kieffer (Eds.), *Handbook of Technology and Literacy.* Mahwah, NJ: Erlbaum.

Risko, V., & Kinzer, C. 1999. *Student Guide for Multi-Media Cases in Reading Education.* New York: McGraw-Hill. (Includes a CD-ROM with yearlong primary reading instruction cases.)

Shulman, L. 1992. Toward a pedagogy of cases. Pp. 1-33 in J. H. Shulman (Ed.), *Case Methods in Teacher Education.* New York: Teachers College Press.

Sykes, G., & Bird, T. 1992. Teacher education and the case idea. *Review of Research in Education, 18,* 457-521.

Basic reviews of the knowledge base relevant to this chapter can be found in the following recent publications:

Kamil, M. L., Mosenthal, P. B., Pearson, P. D., & Barr, R. (Eds.). 2000. *Handbook of Reading Research: Volume III.* Mahwah, NJ: Erlbaum. (See especially Guthrie and Wigfield on engagement and motivation, Purcell-Gates on family literacy, and Gadsden on intergenerational literacy.)

National Research Council. 1998. *Preventing Reading Difficulties in Young Children.* Committee on the Prevention of Reading Difficulties in Young Children, C. E. Snow, M. S. Burns, and P. Griffin (Eds.). Washington, DC: National Academy Press. (See especially Parts II and III.)

Neuman, S. B., & Dickinson, D. K. 2001. *Handbook of Early Literacy Research.* New York: Guilford Press. (See especially Wasik, Dobbins, and Herrmann on intergenerational literacy, Goldenberg on schools working with low-income families, Dickinson and Sprague as well as Roskos and Neuman with different perspectives on the impact of early care environment differences.)

5

Anticipating Challenges

Assessment, Prevention, and Intervention

Today's elementary school teachers can expect their jobs to be more challenging than ever. Students placed at risk of educational failure represent the fastest-growing segment of our school population. Children with varying degrees of learning disabilities receive instruction in regular classrooms under inclusion policies. Large percentages of children continue to live in poverty. Many children attend low-performing schools in poor neighborhoods. More students in American schools are learning English as a new language.

From infancy through age 5, children should not be characterized as having "reading problems," but preschool teachers and day care providers can and must be powerful agents in identifying and helping those youngsters who are *most likely to experience* difficulties in the early primary grades. To do this, they must know how to observe language delays or hearing problems. They must keep their eyes out for the preschooler who still doesn't understand how books "work" or the kindergartner who doesn't recognize any letters. Teachers must move swiftly to arrange for enriched environments—stimulation for oral-language development, with books, songs, story times, phonemic awareness, emergent writing, and other pro-literacy activities. These sorts of prevention efforts can be absolutely critical in changing a child's education future.

Once children are in elementary school, the best prevention is a well-conceived curriculum and good teaching in the primary grades. If children are identified as slipping behind or worse, failing in reading, they need interventions that quickly assess and zero in on their reading problems.

The essential resource for the child is the regular classroom teacher. Every elementary school teacher in America should know how to assess reading problems

Learning on the Job

Mrs. Ryan is the reading specialist in the school where Marie teaches third grade. In her second year of teaching, Marie is full of enthusiasm for the job. She spends long hours preparing for class time and takes special pride in her many efforts to do well, such as bringing good literature into the classroom and assigning innovative writing projects. But still, she is anguished by several students in her class who are not doing well, despite her best efforts.

Marie glances through several students' folders as she waits to speak with Mrs. Ryan. Kevin is far behind his classmates in both reading and writing. Marie notices a comment she wrote after a recent reading lesson: "Kevin has so much difficulty with vocabulary that he loses control of the meaning of the story." Leafing through his papers, she notes that his writing samples look more like those of a beginning first grader than a third grader. His struggle to get one or two sentences down is painfully obvious; almost all of the words are misspelled with few vowels included. Yet Kevin, like many of the others who are having difficulty, seems to be a smart, savvy kid. Kevin is a wonderful conversationalist who tells great jokes. He is interested in science.

Marie admits that Kevin and the other struggling students like him are enigmas to her. There is Julie, who is smart and eager to learn but struggles tremendously because, despite constant practice, she still can't remember the sounds the letters make when she sees them in new words. Iliana, a recent immigrant, does well in math and has rapidly gained an understanding of spoken English but still doesn't speak very much. When she "sounds out" a word, too often she has nothing to match it to in her English vocabulary. Then there is Riki, who has a language disability and participates in an inclusion program, spending part of the day in the resource room and another part with Marie for language arts instruction, but Marie isn't sure it's working well enough.

and how to arrange swiftly for the right kind of supplemental help, usually with a school's reading specialist.

It is essential for the classroom teacher to stay involved. Effective intervention for reading difficulty is not a matter of isolated occasional sessions between the child and a clinician. Instruction in the regular class must be tailored to support and sustain other intervention efforts, collaborating with specialists. Today children often attend after-school and summer programs that feature reading. A child may go to a private speech or language therapist who provides instruction relevant to reading. It's up to the classroom teacher to forge the links with the child's out-of-school life, communicating and collaborating with the child's family, so reading improvement can be attained and sustained.

prepared to know when to intervene

For certain children, particularly those faced with multiple health and socioeconomic challenges from birth, it seems all too clear that learning to read will be an uphill struggle. Teachers should not wait until a child is in the second or third

How perplexing these children are! No matter how well Marie plans her literacy lessons, they never seem to pay full attention, they are frequently distracted, and they rarely understand the follow-up activities she assigns. As the majority of the class moves forward throughout the year, these children seem to fall further and further behind.

Last week Mrs. Ryan told Marie she needs to organize her class time to offer more individualized instruction, but this is easier said than done. Marie tried to focus in on small groups and individuals while the rest of the class silently practiced reading, but many times her "silent" readers were anything but, growing boisterous and out of control.

Today, Marie plans to ask Mrs. Ryan for more specific suggestions to manage her classroom so that all the children are working productively while she takes time to give individual attention. She also wants advice about the right assessment tools to determine exactly what these struggling students need. (She only learned a few techniques in college, and so far they aren't working.)

Marie knows she is doing the right thing by seeking out Mrs. Ryan's help. To the credit of her school system, Mrs. Ryan, who has a master's degree in reading and 15 years of experience, is available at regular times to meet with teachers, talk to them, share resources, and coach them through the rough parts. It's not just finding out what works that Marie needs; it's a matter of making it work in her own class with these kids. And that's where Mrs. Ryan is always there to help Marie try and try again.

Marie knows she is lucky to have the opportunity to collaborate with Mrs. Ryan and draw on her expertise to tackle tough classroom challenges. Not enough new teachers have that kind of support. Marie knows two classmates from college who are just about ready to give up on the teaching profession because they have been left on their own with a sense of failure and inadequacy.

grade, well behind his or her peers, to intervene. And yet, all too often, this is exactly what happens.

Which children are most likely to be placed at risk of failure? When should "prevention" and "intervention" programs take place?

Colleges of education must ensure that every teacher arrives on the job with an understanding of risk factors, and how risk factors work, in connection with literacy and reading development. Much research has documented the risks associated with early childhood language delays, a family history of reading problems, limited proficiency in English, poor neighborhoods, and failing schools. It's important to note that any single risk factor may not be significant on its own, nor do risk factors condemn to reading failure each child who is faced with them. However, a combination of these factors should act as a warning signal to teachers that a child may need extra help along the way—the earlier, the better.

Colleges of education must prepare teachers to intervene with individual children. When teachers get their first jobs, they are often overwhelmed by the wide variety of reading levels they encounter among students in the same class. They need to be better prepared to deal with the collection of individuals that make up a class. They need practice noticing a child, thinking deeply about that child's learning spurts and instructional needs, and paying attention to how the child changes over time. They need to practice the language needed to think and communicate about children's reading with colleagues, specialists, and families.

WHY INTERVENE EARLY?

A recently convened national panel, the Committee on Minority Representation in Special Education, investigated special education in the United States and reached a compelling conclusion:

> At the core of our study is an observation. . . .There is substantial evidence with regard to both behavior and achievement that early identification and intervention [are] more effective than later identification and intervention. This is true for children of any race or ethnic group, and children with or without an identifiable "within-child" problem [neurobiological factors, for example]. Yet the current special education identification process relies on a "wait-to-fail" principle that both increases the likelihood that children will fail because they do not receive early supports and decreases the effectiveness of supports once they are received. . . .
>
> While this principle applies to *all* students, the impact is likely to be greatest on students from disadvantaged backgrounds because (a) their experience outside the school prepares them less well for the demands of schooling, placing them at greater risk for failure and (b) the resources available to them in general education are more likely to be substandard. Early efforts to identify and intervene with children at risk for later failure will help all children who need additional supports. But we would expect a disproportionately large number of those students to be from disadvantaged backgrounds.
>
> The vision we offer in the report is one in which general and special education services are more tightly integrated; one in which no child is judged by the school to have a learning or emotional disability or to lack exceptional talent until efforts to provide high-quality instructional and behavioral support in the general education context have been tried without success. The "earlier is better" principle applies even before the K-12 years. The more effective we are at curtailing early biological harms and injuries and providing children with the supports for normal cognitive and behavioral development in the earliest years of life, the fewer children will arrive at school at risk for failure.

In light of its study results, the committee made the following recommendation:

> *Teacher Quality:* General education teachers need significantly improved teacher preparation and professional development to prepare them to address the needs of students with significant underachievement or giftedness.

SOURCE: National Research Council. 2002. *Executive Summary: Minority Students in Special and Gifted Education*, M. S. Donovan & C. T. Cross (Eds.). Washington, DC: National Academy Press.

prepared to assess all aspects of skilled reading

Watching students fail at reading is a wrenching experience. Even the best teacher will have that experience from time to time. A dedicated, competent teacher may give individual instruction during class time, arrange for supplemental help during school hours, and enlist family support and tutoring programs and then still see a child continue to fall behind, despite hours of extra work.

As we have called for throughout this book, preservice teachers should learn that skilled reading is a complex process involving many parts:

- Readers are able to decode printed words using sound-spelling connections, and they develop an extensive "sight word" repertoire.
- They are able to use prior knowledge, vocabulary, and cognitive strategies to comprehend literal and inferred meanings.
- They read with fluency, swiftly and accurately coordinating word identification with powerful comprehension, so that they read with ease and respond readily to the text.

Children fit lessons into complex backgrounds and daily lives, working their sometimes quirky and always busy minds to put it all together for growing up. Teachers must gather information in order to plan lessons that have a good chance to work with their ever-changing children and then, during lessons, gather more information to determine if the lesson is working for the particular children in it.

Teachers must be ready to examine materials and curriculum mandates in light of what a child is likely to bring to a learning task. They must be practiced at scrutinizing the bases for children's questions and answers. They must learn both to react on the spot and to reflect after the fact on the day's lesson in order to design follow-up lessons that clarify tasks and redirect any children headed down frustrating paths.

Such knowledge is essential to plan the year and the days that make it up, to choose and tailor materials and activities to fit the students. Teachers must be ready to find the ways out of blind alleys and through misconceptions that students may encounter.

When a child is struggling, a good teacher must take steps to find out exactly where the process has gone astray so that he or she can give the right sort of instruction and the materials that will prompt effective practice. It is important to note that individualized instruction does not necessarily mean giving the struggling child one-on-one help with the same lesson that the class is working on. On the contrary, it often means giving the child a different lesson.

What lesson should that be? That depends. Does the student have difficulty comprehending what he or she has read? Does the problem diminish if the text is about certain topics? Does prior access to a table of contents and glossary lead to more success? Does the child recall the details but fail to put ideas together to get the main point? Do directions to ask questions, make summaries, or draw pictures help the child understand a passage?

Does the child have a decoding problem? Does it show in basic words like SAT and PAN or just in multisyllable words like SATIN or PANDEMONIUM? Is there more difficulty with consonant blends at the ends of words? Do vowel digraphs produce uncertainty? Are common words that have odd spellings the ones that stump the child?

Is it a fluency problem? Does the child take a long time to get through a paragraph and then find it hard to remember enough to pull together the meaning and give a response to it? Does the child read well-known passages with comfort but stumble and complain when given new ones?

The only way to know is through good assessment techniques. And yet most of today's elementary school teachers don't have sufficient time or expertise to assess children effectively. As a result, many schools leave assessments in the hands of reading specialists, usually called on after a child is already behind or failing. That's not good enough.

We need to prepare classroom teachers to see more clearly and understand more deeply the learners in their charge. This is crucial not just for struggling students but also for children who accelerate ahead of grade level and need more challenging tasks.

Knowing how to zero in on the unique strengths, weaknesses, and needs is a career-long and complex challenge. But at the preservice level, teachers-to-be should have hands-on experiences collecting data about individual children and contributing to decisions about their instruction. To be effective, they must learn not just a few assessment techniques but many—ranging from simply being a good observer of children to techniques for using more formal standardized tools.

Teachers should be comfortable administering and interpreting norm-referenced and criterion-referenced instruments to determine where their students are as readers. A general test-and-measurement course is helpful but not sufficient. The fate of struggling readers depends on teachers understanding the specifics of reading assessment. Finally, all teachers should be familiar with state and local standards and curriculum guides because of the implications for assessing student progress during the course of the school year.

prepared to provide interventions

Recent national studies and reports have shown a growing body of evidence regarding the most effective practices for teaching reading. However, many of these approaches have not gone the distance from university researchers to the field, nor to enough of the nation's colleges of education. This is also true of intervention programs designed for struggling readers.

The nature, tone, and delivery of effective reading intervention programs vary greatly. Teachers should be aware that, according to research, successful intervention programs share certain essential elements:

- The programs give failing readers and writers more time on task and feature materials that students will enjoy and can handle successfully.
- They use a variety of activities, including rereading familiar texts, introduction of new texts, writing opportunities, and word study, including systematic phonics.
- Good interventions frequently monitor individual progress and seek the involvement of families.
- Extra training and support for aides, volunteers, and teachers characterize most effective practices.

Districts and schools should help teachers hone their skills as good managers and coordinators of the many players involved in interventions: speech therapists, reading specialists, volunteer tutors, social workers, and families.

Reading interventions take many shapes and forms. When people say "intervention" they often think only of pull-out "supplemental" or "remedial" programs, where a reading specialist works with individual or small groups of children who are having difficulties learning to read. But for children struggling to learn to read, effective intervention must take many guises, not just a single supplemental program. In fact, the term should refer to a coordinated array of efforts.

The regular classroom teacher must manage a collaboration with supplemental programs. It is up to the teacher to provide instruction during class time that is well designed to support the lessons given by specialists.

Interventions may involve support from psychological, social, and health professionals, in or out of school. Besides that, paraprofessionals and volunteers are often assigned to intervene with struggling readers. Intensified help may take the form of a tutor in an after-school program. An aide may assist a child identified for special education who is served by inclusion in a multiple-abilities classroom.

Unless each encounter is well planned by a professional who is knowledgeable about reading instruction, and unless the aides, tutors, or volunteers are well prepared, coached, and supervised (often by a reading specialist), the time and money expended may have no payoff in terms of improving reading. But when well done and as part of a coordinated effort, they can be helpful.

Classroom teachers extend the power of the intervention as they guide and advise parents on how to help struggling children with homework, practice, and good literature. Families are key to sustained and effective intervention. Teachers need to learn from families about interventions that have been tried in the past but do not show up in the official records. Too many children have experienced a series of unconnected efforts that promised much and yielded little. A good intervention effort needs to confront and counter the cumulative effect that such a history has on the child and the family.

School systems should offer support and opportunities for teachers to work collaboratively with families and see them as partners in success. In Chapter 4 there is an extensive discussion about teachers connecting home and school efforts. For children experiencing reading difficulties, two way communication and coordination are doubly—maybe 10-fold—important.

Unless someone is there to monitor and act as a communications liaison, interventions may become fragmented, inconsistent, contradictory, and even ineffective. In all cases the regular classroom teacher is in the best position to be a powerful force in rallying and coordinating all the agents at work.

instructional groupings in interventions

Teachers need preservice and ongoing professional development based on recent research that helps them make well-informed decisions about the programs and materials they choose. New approaches, however, may stretch a teacher's classroom management skills. The standard class organization plan and the tried-and-true behavior management techniques may not fit with the best approach for struggling students.

It is difficult for a teacher to gear instruction to struggling readers if the classroom is always managed in a whole-group manner. If children do not get instruction tailored to their needs, it is highly unlikely they will improve on their own. The whole group will move further and further ahead, while slower readers are left behind, still struggling with what the others take for granted.

All teachers should know how to manage their classrooms so that they can provide instruction to different configurations of students, depending on the activity. Children may be assigned to groups that have the same range of abilities (homogeneous groups) or to groups that have a mix of ability levels (heterogeneous groups).

Sessions with the whole class call for preparation and moment-to-moment teacher actions that are quite specialized. A different strategy and tactics are needed to teach one small group while the other students are less directly supervised. A good teacher can accomplish lesson goals with a small group while coordinating a class with some students working in pairs or small peer groups and others working on their own.

computers in interventions

There are wonderful new technology applications for teaching. Computers are alluring to some students who would otherwise resist reading and writing. Computers can be excellent tools for testing and assessing students. Computers can also offer students valuable time to practice and apply what they have been taught.

For some the use of computers in reading instruction is associated with an image of worksheets—using monitors and keyboards instead of scratchy paper and number two pencils. But now there are computer applications with interesting stories and informational text that can provide a scaffold for a child's developing abilities in comprehension, vocabulary, word identification, and fluency. Thanks to advanced hardware capabilities, some specialized software can even assist with correcting read-alouds, sounding out a new word, or stimulating the use of phonics or metacognitive strategies that have been introduced in the curriculum. In addition, there are special uses of more general computer applications—word processing, database, or publishing software—that can support reading and writing instruction. Collaborative projects using the Internet provide practice in reading and writing across the curriculum from science to geography and more. But computers have their limitations, and teachers must avoid mindless practice devoid of challenge or purpose.

Teacher education and professional development should show how technology can be used to support and reinforce larger curricular goals and the individual needs of students. Time spent on a computer can and should be fun. But good teachers use technology for specific, rather than open-ended, purposes, with direct connections to the classroom reading instruction.

Because technology is constantly changing, school districts are in the best position to help keep teachers up to date with the most current software and learning applications. Onsite technology coaches should be routinely available

Bernadine Hansen, District Literacy Coordinator, Tyler Independent School District, Tyler, Texas

Coach the Teacher, Help the Children

I coach 10 teachers, kindergarten through second grade. I give them classes at night—2 hours every other week, all year long. Then I am available on the elementary school campus for further direction, information, and guidance.

In most other places, in-service training happens on a surface level. The teachers go back to their classrooms with one or two ideas to implement. Maybe the ideas don't work, or maybe they do but don't go much further. My whole purpose is to help teachers understand the theory behind the applications. That's what we lack in our professional development and colleges of education.

It's a collaboration. I have a teacher's salary. I get no extra pay. I'm not an administrator. I'm not an evaluator. I work side by side with the teachers. I do a lot of modeling for teachers, showing them how to use techniques, like shared writing experiences, in their classrooms.

So many teachers are out there on their own. They're in a four-wall cell, and they don't get to communicate with one another. They're handed down mandates that children must read at a certain level. But they often lack the theory of reading and writing they need to carry this out. Maybe they have six kids in their class who can't read or write, and they are out there on a limb by themselves with no support. So the teachers really value having an expert to talk to—someone who knows something about the reading and writing process, a coach, right there on campus to consult with on a daily basis about children who are having difficulty.

We use diagnostic tools to teach us where to go—like running record and independent writing samples. We go from what the child knows to what the child doesn't know. We always look at the positive. What does the child have in place and where are we going to help that child next?

and can and should lead infrequent but regular technology workshops to guide and support teachers.

The reading coach and the technology specialist must collaborate to help classroom teachers assess the array of possibilities that new technologies offer. Each new application has to be carefully considered. Does it fit the curriculum? Is it attuned to the goals and processes of reading instruction? Has it been shown to contribute to reading improvement? What grade and reading level is it good for? Where and when can it fit the school and class routines so that it has the best chance to help children? How can the teacher tell when it is working and when it needs to be changed to help a particular student?

In one important way the new information technologies are not all that different from any aspect of teaching reading. Everyone involved must keep their eyes on the prize; it is all about a particular child improving his or her reading power.

professional development and assessment

Teachers get a different view of the learning they work with every day when it is filtered through assessment procedures and analyses. They notice new things about the children and the academic subject, and they see relationships they hadn't had a chance to notice before. Consulting with a peer or an expert about assessment results can be a form of professional development itself. But there are other links between assessment and professional development that make it an apt subject to close our book.

First, there are concerns about the role of assessment in the future of reading instruction. Will the prevalence of some types of assessment lead to a narrowing of the curriculum? Will we have a pseudosuccess if children learn to pass tests but the time taken to prepare for them robs children of some of the opportunities they need to become skilled readers? Well-prepared teachers will alert us when such problems appear.

There have been projects with researchers and teachers aimed at producing assessment instruments that would be unlikely to narrow the curriculum and produce pseudosuccesses. These are research and development projects. But a funny thing happens. Time and again, the participants notice that the work sessions are, in fact, high-quality professional development. Trying to design an assessment instrument or procedure leads to deep understandings about the instruction of reading as well as strong dispositions to improve instruction.

There is another important link between assessment and professional development. The consensus of researchers and practitioners is that assessment results are a cornerstone for high-quality professional development. When professional

QUALITY PROFESSIONAL DEVELOPMENT FOR TEACHERS

The National Staff Development Council has produced a set of standards that reflect the research, policy, and practitioner consensus about the features that bring quality to professional development for teachers:

Context standards—Staff development that improves the learning of all students:

- Organizes adults into learning communities whose goals are aligned with those of the school and district (Learning Communities).
- Requires skillful school and district leaders who guide continuous instructional improvement (Leadership).
- Requires resources to support adult learning and collaboration (Resources).

Process standards—Staff development that improves the learning of all students:

- Uses disaggregated student data to determine adult learning priorities, monitor progress, and help sustain continuous improvement (Data-Driven).
- Uses multiple sources of information to guide improvement and demonstrate its impact (Evaluation).
- Prepares educators to apply research to decisionmaking (Research-Based).
- Uses learning strategies appropriate to the intended goal (Design).
- Applies knowledge about human learning and change (Learning).
- Provides educators with the knowledge and skills to collaborate (Collaboration).

Content Standards—Staff development that improves the learning of all students:

- Prepares educators to understand and appreciate all students; create safe, orderly, and supportive learning environments; and hold high expectations for their academic achievement (Equity).
- Deepens educators' content knowledge, provides them with research-based instructional strategies to assist students in meeting rigorous academic standards, and prepares them to use various types of classroom assessments appropriately (Quality Teaching).
- Provides educators with knowledge and skills to involve families and other stakeholders appropriately (Family Involvement).

SOURCE: National Staff Development Council. 2001. *NSDC Standards for Staff Development.* Oxford, OH: Author.

development addresses the specific content that assessment results show a teacher needs to work on in order to improve his or her own students' achievement, it is valued by the teacher and is more likely to have an impact.

We and this book have a very small role in the whole endeavor of teacher education and professional development for improving reading instruction. We focus on the content knowledge for providing children the opportunities they need to become skilled readers. The final piece of content is about the research that content knowledge is based on.

Regarding research, teachers are the assessors. Teachers should be given opportunities to be involved with research—by keeping track of developments, evaluating new findings, considering how it matters to what they are doing and could do with their students, contributing to the knowledge base, and spreading the word about it. It is true that we have learned a great deal about beginning reading. It is also true that we have the opportunity and the need to build on it and that teachers are as important to future research as they are to the futures of their students.

Headlines say that teachers are expected to be *accountable to* parents, principals, school boards, and taxpayers. Teachers know this is a minuscule burden compared to the real challenge: They are *responsible for* the children. We echo teachers who ask for professional development in order to fulfill that responsibility.

Tinesha now knows how to write and read her heart's desire: I want my family to be happy.

Activity

One of the Class

Even before internship or student teaching, preservice teacher education can prepare students to trace how children change over time and to recognize that any given elementary school classroom is likely to have children with diverse abilities. In this activity, spanning a full semester course, each teacher education student will trace a child throughout a simulation of the school year. (The activity is a small addition to the regular instruction in a course for which the materials described below are relevant.)

The preparation of materials for this activity begins a year before it is used for teacher education. A faculty member enlists the help of a classroom teacher and visits regularly to assemble records of children's reading and writing performance throughout the year. By June there are writing samples, audiotapes of reading, teacher-made tests, and results of external testing.

- Give each teacher education student responsibility to trace one child in the elementary school class. Put the materials on library or laboratory reserve, sorted by period:

Period 1	School opening until the end of September
Period 2	From October until the end of December
Period 3	From January until the end of the school year

- Early in the course, introduce the materials collected from Period 1. Tell each student to look at or listen to all materials relevant for "your" elementary school child. In class, examine the children's samples in the context of the group of children. Discuss the wide range of performance levels evident in the elementary classroom. Guide the students in using a rubric (scoring guide) to analyze their children's materials for strengths and potential teaching points.

- About a month later introduce materials from Period 2. Again begin with a discussion of the diversity in the elementary classroom. Provide a rubric and have the students analyze the new material. Have the students look carefully at their children's changes over time, preparing a letter for the family that explains where progress is being made and where weaknesses persist. Critique their letters for accuracy and appropriateness.

About a month later distribute the rest of the materials. Have the students focus on the yearlong growth of the elementary student each is responsible for covering. Ask them to prepare a report for the reading specialist, recommending an instructional plan for the child's next year. Critique their reports for accuracy and appropriateness.

Activity

What Does the Child Need?

In-service teachers or interns often need to sharpen their skills for recognizing a reader's strengths and weaknesses. This activity begins with group analysis of an adult simulation and moves to a careful critique of work with a child. Finally, the effort is related to the instruments that are mandated or available in the schools the teachers work in. There should be an experienced reading specialist as a leader for the activity.

1. Select a piece of dense text from a publication such as *Science* magazine or a computer programming manual. Be sure to select an excerpt that includes little-known technical terms, difficult-to-pronounce words, and challenging content. Make copies and distribute to the class, and then select a student who will stand up before the class and read aloud. (Select a student who can withstand the attention and set a tone that will avoid embarrassment.) Ask his or her classmates to do an assessment to figure out what problems the reader has. Ask them to note which words, exactly, are difficult to decode and why. At what point does the reader seem to lose the meaning? Which vocabulary words are baffling? This part of the activity should provide perspective on assessment and sensitivity to the many ways in which the reading process can go awry.

2. Next, ask each student to work with a child in the classrooms where they are teachers or student teachers. Help them choose an appropriately challenging piece of text for the child. Have them make a tape recording of the child reading the passage aloud and answering questions about its meaning. Have them transcribe the tape. Select a transcript to use with the entire class. Have the class read the transcript (and the original passage). Review the aspects of skilled reading and what can go awry (see pp. 151-152). Notice patterns of errors and evidence of strength. Have the class work together to suggest and discuss exactly what sorts of materials and instructions should be tailored to the child's individual needs. Have each student analyze his or her own transcript and ask for advice where needed.

3. Next, assign small groups to study and report to the class about different informal and formal assessment techniques. In response to each presentation, lead the group to consider how combinations of assessment instruments and techniques serve to help teachers decide about grouping and individual instruction.

Activity

Finding the Intervention

In a study group led by a reading specialist, teachers can increase their understanding of the features of effective interventions for reading. In the process, they will collaborate to produce a resource for local parents and teachers.

1. Devise a checklist to describe approaches to intervention for children experiencing difficulty learning to read. Include the features on p. 153. Make sure everyone has the same working definition of each item on the checklist by working on hypothetical examples. Add items for basic information about age, grade level, cost, timing, entry and exit criteria, evidence of effectiveness, and so forth.

2. As homework, gather lists of intervention programs and approaches for reading that are available in the local school district. Interview curriculum specialists, principals, and reading specialists. Check the phone book for private services. Ask about efforts by civic or church groups.

3. Divide up the list and divide the group into teams, so that each program or approach on the list is covered. The team should visit, interview, observe, and collect materials; it should do whatever it can to apply the checklist and describe each intervention on the list.

4. Discuss the descriptions and revise for clarity. Make categories in order to group the intervention approaches and programs in a directory.

5. Post the directory on the World Wide Web or make it available through a library. Include directions for people who will want to update it in the future.

Activity

Learning to Manage via Cross-Classroom Visits

Managing small groups is a challenge. Some teachers lose focus when more than one thing is going on at once. If a teacher is finding that small-group time means pandemonium, a "cross-classroom visitation" can be arranged to maximize learning from a nearby teacher who excels in classroom management of small reading groups.

The school's reading specialist or a district curriculum coordinator is a key player in this professional development activity. Novice or resident teachers first visit an exemplary classroom for targeted observations and are later visited by coaches who help them implement the new practice, so that they, too, will soon be exemplary! To make this chain of events work well, the reading specialist or curriculum coordinator should take the following steps:

- Before the observation, spend time with the teacher who wants help in order to discuss how and why small groups work and to give some tips about what to look for during the visit to the expert's classroom. Further prepare the observer by examining the assessment procedures the teacher relied on to assign students to different groups.

- After the observation, meet again for a debriefing. Listen carefully to the problems the observing teacher noticed as well as to any problems he or she suspects might arise during an attempt to replicate what was observed. Provide practice for parts of the procedure; help plan a gradual phase-in.

- Visit the teacher during the reading period to coach and carry out the phase-in plan. Find out which kids are having trouble with comprehension strategies, such as following the plot, understanding the concept of a narrator, or identifying genres. Organize those children in a small group and provide a 10- or 15-minute lesson. Do the same with another group that is having decoding problems. Let the teacher observe while you work with his or her students in this way, while keeping the rest of the class well focused on other activities, such as silent reading or writing. At times, roam the class to provide individual attention to one or two students. Fit in a whole-group lesson—all during one language arts block.

- On succeeding days, hand over parts of the teaching to the classroom teacher. Stay nearby to render assistance but begin to focus on consultation and providing assistance to the teacher in assessing students for group placement.

- Six weeks after the first cross-classroom visitation, meet with the teacher again to find out if it was helpful and to answer any questions. If necessary, enter another coaching phase.

resources

Basic reviews of the knowledge base relevant to this chapter can be found in the following recent publications:

Kamil, M. L., Mosenthal, P. B., Pearson, P. D., & Barr, R. (Eds.). 2000. *Handbook of Reading Research: Volume III.* Mahwah, NJ: Erlbaum. (See especially Shaywitz and colleagues on the neurobiology of dyslexia; Klenk and Kibby on remediation; Hiebert and Taylor on early intervention; and Anders, Hoffmann, and Duffy on teaching teachers.)

National Research Council. 1998. *Preventing Reading Difficulties in Young Children,* Committee on the Prevention of Reading Difficulties in Young Children. C. E. Snow, M. S. Burns, and P. Griffin (Eds.). Washington, DC: National Academy Press. (See especially Parts II and III, Chapters 5 and 8).

Neuman, S. B., & Dickinson, D. K. (Eds.). 2001. *Handbook of Early Literacy Research.* New York: Guilford Press. (See especially Barnett on the effects of preschool as a prevention/ intervention, Invernizzi on tutoring, McGill-Franzen and Goatley on Title I and special education, Johnston and Rogers on informed assessment, Salinger on assessment with multiple forms of evidence, Watson on implications of oral-language development, Scarborough on connections between early language and subsequent literacy, Vellutino and Scanlon on early interventions and individual differences, and Strickland on African American children considered to be at risk.)

Prepared to know when to intervene so children are not placed at more risk, examples of overviews and metanalyses:

Clay, M. M. 1989. *The Early Detection of Reading Difficulties, 3rd ed.* Hong Kong: Heinemann.

García, G. E. 1991. Factors influencing the English reading test performance of Spanish-speaking Hispanic children. *Reading Research Quarterly, 26,* 371-392.

La Paro, K. M., & Pianta, R. C. 2000. Predicting children's competence in the early school years: A meta-analytic review. *Review of Educational Research, 70*(4), 443-484.

Leseman, P. P. M., & de Jong, P. F. 1998. Home literacy: Opportunity, instruction, cooperation and social-emotional quality predicting early reading achievement. *Reading Research Quarterly, 33,* 294-318.

Lombardino, L. J., Riccio, C. A., Hynd, G. W., & Pinheiro, S. B. 1997. Linguistic deficits in children with reading disabilities. *American Journal of Speech-Language Pathology, 6,* 71-78.

Pellegrini, A. D. 1991. A critique of the concept of at risk as applied to emergent literacy. *Language Arts, 68,* 380-385.

Scarborough, H. S. 1998. Early identification of children at risk for reading disabilities: Phonological awareness and some other promising predictors. Pp. 77-121 in B. K. Shapiro, P. J. Accardo, & A. J. Capute (Eds.), *Specific Reading Disability: A View of the Spectrum.* Timonium, MD: York Press.

Scarborough, H. S., Dobrich, W., & Hager, M. 1991. Preschool literacy experience and later reading achievement. *Journal of Learning Disabilities, 24,* 508-511.

Share, D., Jorm, A., Maclean, R., & Matthews, R. 1984. Sources of individual differences in reading acquisition. *Journal of Educational Psychology, 76,* 1309-1324.

Teisl, J. T., Mazzocco, M. M. M., & Myers, G. F. 2001. The utility of kindergarten teacher ratings for predicting low academic achievement in first grade. *Journal of Learning Disabilities, 34,* 286-293.

Torgesen, J. K. 1995. *Phonological Awareness: A Critical Factor in Dyslexia.* Baltimore: The Orton Dyslexia Society.

Prepared to know when to intervene, examples of more specific studies and reports:

Mantzicopoulos, P. Y., & Morrison, D. 1994. Early prediction of reading achievement: Exploring the relationship of cognitive and noncognitive measures to inaccurate classifications of at-risk status. *Remedial and Special Education, 15,* 244-251.

Normandeau, S., & Guay, F. 1998. Preschool behavior and first-grade school achievement: The mediational role of cognitive self-control. *Journal of Educational Psychology, 90,* 111-121.

Reese, L., Garnier, H., Gallimore, R., & Goldenberg, C. 2000. Longitudinal analysis of the antecedents of emergent Spanish literacy and middle-school English reading achievement of Spanish-speaking students. *American Educational Research Journal, 37,* 633-662.

Shankweiler, D., Lundquist, E., Katz, L., Stuebing, K. K., Fletcher, J. M., Brady, S., Fowler, A., Dreyer, L. G., Marchione, K. E., Shaywitz, S. E., & Shaywitz, B. A. 1999. Comprehension and decoding: Patterns of association in children with reading difficulties. *Scientific Studies of Reading, 31,* 24-53, 69-94.

Signorini, A. 1997. Word reading in Spanish: A comparison between skilled and less skilled beginning readers. *Applied Psycholinguistics, 18,* 319-344.

Stanovich, K. E., & Siegel, L. C. 1994. Phenotypic performance profile of children with reading disabilities: A regression-based test of the phonological-core variable-difference model. *Journal of Educational Psychology, 86,* 24-53.

Torgesen, J. K. 2000. Individual responses to early interventions in reading: The lingering problem of treatment resisters. *Learning Disabilities Research & Practice, 15,* 55-64.

Vellutino, F. R. 2001. Further analysis of the relationship between reading achievement and intelligence: Response to Naglieri. *Journal of Learning Disabilities, 34,* 306-310.

Williams, J. P. 1993. Comprehension of students with and without learning disabilities: Identification of narrative themes and idiosyncratic text representations. *Journal of Educational Psychology, 85,* 631-641.

Assessment, commentaries and overviews:

American Federation of Teachers, National Council on Measurement in Education, and National Education Association. 1990. *Standards for Teacher Competence in Educational Assessment of Students.* Washington, DC: American Federation of Teachers.

Black, P., & William, D. 1998. Inside the black box. *Phi Delta Kappan, 80*(2), 139-147.

Cooper, J. D., & Kiger, N. 2001. *Literacy Assessment.* New York: Houghton Mifflin.

Glazer, S. M. 1998. *Assessment Is Instruction: Reading, Writing, Spelling, and Phonics for All Learners.* Norwood, MA: Christopher-Gordon.

Goodwin, W. L., & Goodwin, L. D. 1997. Using standardized measures for evaluating young children's learning. Pp. 92-107 in B. Spodek & O. Saracho (Eds.), *Issues in Early Childhood Educational Assessment and Evaluation.* New York: Teachers College Press.

Hargreaves, A., Earl, L., & Schmidt, M. 2002. Perspectives on alternative assessment reform. *American Educational Research Journal, 39*(1), 69-95.

International Reading Association. 1999. High-stakes assessments in reading: A position statement of the International Reading Association. *Reading Teacher, 53,* 257-263.

International Reading Association/National Council of Teachers of English Joint Task Force on Assessment. 1994. *Standards for the Assessment of Reading and Writing.* Newark, DE: Author.

Linn, R. 2000. Assessments and accountability. *Educational Researcher, 29*(2), 4-16.

Madaus, G. F. 1999. The influence of testing on the curriculum. Pp. 73-111 in M. M. Early & K. J. Rehage (Eds.), *Ninety-Eighth Yearbook of the National Society for the Study of Education.* Chicago: University of Chicago Press.

Popham, J. 1995. *Classroom Assessment: What Teachers Need to Know.* Needham Heights, MA: Allyn & Bacon.

Torgesen, J. K., & Wagner, R. K. 1998. Alternative diagnostic approaches for specific developmental reading disabilities. *Learning Disabilities Research and Practice, 13,* 220-232.

Winograd, P., & Arrington, H. J. 1999. Best practices in literacy assessment. Pp. 210-241 in L. B. Gambrell, L. M. Morrow, S. B. Neuman, & M. Pressley (Eds.), *Best Practices in Literacy Instruction.* New York: Guilford.

Assessment, examples of more specific studies, reports, and instruments:

Cairns, H., McDaniel, D., & McKee, C. (Eds.). 1996. *Methods for Assessing Children's Syntax.* Cambridge, MA: MIT Press.

Harp, B., & Brewer, J. 2000. Assessing reading and writing in the early years. Pp. 154-167 in D. S. Strickland and L. M. Morrow (Eds.), *Beginning Reading and Writing.* New York: Teachers College Press and the International Reading Association.

Hresko, W. P., Peak, P. K., Herron, S. R., & Bridges, D. L. 2000. *Young Children's Achievement Test (YCAT): A Measure to Help Identify Preschool, Kindergarten, and First Grade Children Who Are at Risk for School Failure.* Austin, TX: Pro-Ed.

Kapinus, B. A. 2002. Assessment of reading programs. Pp. 118-128 in S. B. Wepner, D. S. Strickland, J. T. Feeley (Eds.), *The Administration and Supervision of Reading Programs.* New York: Teachers College Press.

McMillan, J. H. 2001. *Experts in Assessment: Essential Assessment Concepts for Teachers and Administrators.* Thousand Oaks, CA: Corwin Press.

National Research Council. 2000. Assessment in early childhood education. Pp. 233-260 in *Eager to Learn: Educating Our Preschoolers.* Washington, DC: National Academy Press.

Paris, S. 2001. *How Can I Assess Children's Early Reading Achievement?* Ann Arbor, MI: Center for the Improvement of Early Reading Achievement.

Plake, B. S., & Impara, J. C. 1997. Teacher assessment literacy: What do teachers know about assessment? Pp. 55-68 in G. D. Phye (Ed.), *Handbook of Classroom Assessment: Learning, Adjusting and Achieving.* New York: Academic Press.

Stecker, P. M., & Fuchs, L. S. 2000. Effecting superior achievement using curriculum-based measurement: The importance of individual progress monitoring. *Learning Disabilities Research and Practice, 15,* 128-134.

Torgesen, J. K., & Wagner, R. K. 1999. *Comprehensive Tests of Phonological Processes.* Austin, TX: Pro-Ed.

Yopp, H. K. 1995. A test for assessing phonemic awareness in young children. *The Reading Teacher, 49,* 20-29.

Preventing and intervening, examples of overviews, metanalyses, and comparisons:

Elbaum, B., Vaughn, S., Hughes, M. T., &. Moody, S. W. 2000. How effective are one-to-one tutoring programs in reading for elementary students at-risk for reading failure? A meta-analysis of the intervention research. *Journal of Educational Psychology, 92,* 605-619.

Johnston, P., & Allington, R. L. 1991. Remediation. Pp. 984-1012 in R. Barr, M. L. Kamil, P. B. Mosenthal, & P. D. Pearson (Eds.), *Handbook of Reading Research, vol. 2.* White Plains, NY: Longman.

McKenna, M. C. In press. *Help for Struggling Readers: Strategies for Grades 3-8.* New York: Guilford.

Pikulski, J. J. 1994. Preventing reading failure: A review of five effective programs. *The Reading Teacher, 48,* 30-39.

Pinnell, G. S., Lyons, C. A., DeFord, D. E., Bryk, A., & Seltzer, M. 1994. Comparing instructional models for the literacy education of high risk first graders. *Reading Research Quarterly, 29,* 8-39.

Strickland, D. S., Ganske, K., & Monroe, J. K. 2002. *Supporting Struggling Readers and Writers: Strategies for Classroom Intervention 3-6.* Portland, ME: Stenhouse.

Swanson, H. L. 1999. Reading research intervention for students with LD: A meta-analysis of intervention outcomes. *Journal of Learning Disabilities, 32,* 504-532.

Torgesen, J. K., Wagner, R. K., & Rashotte, C. A. 1997. Prevention and remediation of severe reading disabilities: Keeping the end in mind. *Scientific Studies of Reading, 1,* 217-234.

Troia, G. A. 1999. Phonological awareness intervention research: A critical review of the experimental methodology. *Reading Research Quarterly, 34,* 28-52.

Preventing and intervening, examples of more specific studies and reports:

Baker, S., Gersten, R., & Keating, T. 2000. When less may be more: A 2-year longitudinal evaluation of a volunteer tutoring program requiring minimal training. *Reading Research Quarterly, 35,* 494-519.

Blachman, B. A., Tangel, D. M., Ball, E. W., Black, R., & McGraw, C. K. 1999. Developing phonological awareness and word recognition skills: A two year intervention with low-income, inner-city children. *Reading and Writing, 11,* 239-273.

Boyle, J. R., & Weishaar, M. 1997. The effects of expert-generated versus student-generated cognitive organizers on the reading comprehension of students with learning disabilities. *Learning Disabilities Research and Practice,* 12(4), 228-235.

Brady, S., Fowler, A., Stone, B., & Winbury, N. 1994. Training phonological awareness: A study with inner-city kindergarten children. *Annals of Dyslexia, 44,* 26-59.

Brown, R., Pressley, M., Van Meter, P., & Schuder, T. 1996. A quasi-experimental validation of transactional strategies instruction with low-achieving second-grade readers. *Journal of Educational Psychology, 88*(1), 18-37.

Buntaine, R. L., & Costenbader, V. K. 1997. The effectiveness of a transitional prekindergarten program on later academic achievement. *Psychology in the Schools, 34,* 41-50.

Campbell, F. A., & Ramey, C. T. 1994. Effects of early intervention on intellectual and academic achievement: A follow-up study of children from low-income families. *Child Development, 65,* 684-698.

Echevarria, J., & McDonough, R. 1995. An alternative reading approach: Instructional conversations in a bilingual special education setting. *Learning Disabilities Research and Practice, 10*(2), 108-119.

Foorman, B., Francis, D. J., Fletcher, J. M., Schatschneider, C., & Mehta, P. 1998. The role of instruction in learning to read: Preventing reading failure in at-risk children. *Journal of Educational Psychology, 90,* 37-55.

Fuchs, D., Fuchs, L. S., & Burish, P. 2000. Peer-assisted learning strategies: An evidence-based practice to promote reading achievement. *Learning Disabilities Research and Practice, 15,* 85-91.

Kreuger, E., & Townshend, N. 1997. Reading clubs boost second-language first graders' reading achievement. *Reading Teacher, 51,* 122-127.

Martineu, G., Lamarche, P. A., Marcoux, S., & Bernard, P. M. 2001. The effect of early intervention on academic achievement of hearing-impaired children. *Early Education and Development, 12,* 275-289.

McGuinness, C., McGuinness, D., & McGuinness, G. 1996. Phonographix: A new method of remediating reading difficulties. *Annals of Dyslexia, 46,* 73-96.

Morris, D., Tyner, B., & Perney, J. 2000. Early steps: Replicating the effects of a first-grade reading intervention program. *Journal of Educational Psychology, 92,* 681-693.

Reynolds, A. J., & Temple, J. A. 1998. Extended early childhood intervention and school achievement: Age thirteen findings from the Chicago longitudinal study. *Child Development, 69,* 231-246.

Shanahan, T., & Barr, R. 1995. Reading Recovery: An independent evaluation of the effects of an early instructional intervention for at-risk learners. *Reading Recovery Quarterly, 30,* 958-996.

Slavin, R. E., & Madden, N. 1999. Effects of bilingual and English as a second language adaptations of *Success for All* on the reading achievement of students acquiring English. *Journal of Education for Students Placed at Risk, 4,* 393-416.

Stuart, M. 1999. Getting ready for reading: Early phoneme awareness and phonics teaching improves reading and spelling in inner-city second language learners. *British Journal of Educational Psychology, 69,* 587-605.

Torgesen, J. K., Alexander, A. W., Wagner, R. K., Rashotte, C. A., Voeller, K., Conway, T., & Rose, E. 2001. Intensive remedial instruction for children with severe reading disabilities: Immediate and long-term outcomes from two instructional approaches. *Journal of Learning Disabilities, 34,* 33-58.

Tzuriel, D., Kaniel, S., Kanner, E., & Haywood, H. C. 1999. Effects of the "Bright start" program in kindergarten on transfer and academic achievement. *Early Childhood Research Quarterly, 14,* 111-141.

Wilder, A. A., & Williams, J. P. 2001. Students with severe learning disabilities can learn higher order comprehension skills. *Journal of Educational Psychology, 93,* 268-277.

Williams, J. 1998. Improving the comprehension of disabled readers. *Annals of Dyslexia, 48,* 213-238.

Instructional grouping choices, examples of studies and reports:

Cunningham, P. M., Hall, D. P., & Defee, M. 1998. Nonability-grouped, multilevel instruction: Eight years later. *The Reading Teacher, 51,* 652-654.

Holloway, J. H. 2001. Grouping students for increased achievement. *Educational Leadership, 59*(3), 84-85.

Juel, C. 1990. Effects of reading group assignment on reading development in first and second grade. *Journal of Reading Behavior, 22,* 233-254.

Lou, Y., Abrami, P. C., & Spence, J. C. 2000. Effects of within-class grouping on student achievement: An exploratory model. *Journal of Educational Research, 94,* 101-112.

Lou, Y., Abrami, P. C., Spence, J. C., Paulsen, C., Chambers, B., & Apollonia, S. 1996. Within-class grouping: A meta-analysis. *Review of Educational Research, 66,* 423-458.

McDermott, R., & Varenne, H. 1995. Culture as disability. *Anthropology & Education Quarterly,* 26, 324-348.

Slavin, R. E. 1993. Ability grouping in the middle grades: Achievement effects and alternatives. *Elementary School Journal, 93,* 535-552.

Computers in interventions, examples of studies and reports:

Davidson, J., Elcock, J., & Noyes, P. 1996. A preliminary study of the effect of computer-assisted practice on reading attainment. *Journal of Research in Reading, 19*(2), 102-110.

Greenlee-Moore, M. E., & Smith, L. L. 1996. Interactive computer software: The effects on young children's reading achievement. *Reading Psychology, 17,* 43-64.

Leu, D. J., Jr., 1997. Caity's question: Literacy as deixis on the Internet. *Reading Teacher, 51*(1), 62-67.

Leu, D. J., Jr., & Leu, D. D. 1997. *Teaching with the Internet: Lessons from the classroom.* Norwood, MA: Christopher-Gordon.

Means, B. (Ed.) 1994. *Technology and Education Reform: The Reality Behind the Promise.* San Francisco: Jossey-Bass.

Means, B., & Olson, K. 1997. *Studies of Education Reform: Technology and Education Reform.* Washington, DC: Office of Educational Research and Improvement.

Pinkard, N. 2001. "Rappin' Reader" and "Say Say Oh Playmate": Using children's childhood songs as literacy scaffolds in computer-based learning environments. *Journal of Educational Computing Research, 25*(1):17-34.

Reinking, D., Labbo, L., & McKenna, M. 1997. Navigating the changing landscape of literacy: Current theory and research in computer-based reading and writing. Pp. 77-92 in J. Flood, S. B. Heath, & D. Lapp (Eds.), *Handbook of Research on Teaching Literacy Through the Communicative and Visual Arts.* New York: Macmillan Library Reference.

Sharp, D. L. M., Bransford, J. D., Goldman, S. R., Risko, V. J., Kinzer, C. K., & Vye, N. J. 1995. Dynamic visual support for story comprehension and mental model building by young, at-risk children. *Educational Technology Research and Development, 43*, 25-42.

Professional development and assessment, examples of studies and reports:

DuFour, R. 1997. Functioning as learning communities enables schools to focus on student achievement. *Journal of Staff Development, 18*(2), 56-57.

Hoffman, J., & Pearson, P. D. 2000. Reading teacher education in the next millennium: What your grandmother's teacher didn't know that your granddaughter's teacher should. *Reading Research Quarterly, 35*(1): 28-44.

Noyce, P., Perda, D., & Traver, R. 2000. Creating data-driven schools. *Educational Leadership, 57*(5), 52-56.

Spalding, E. 2000. Performance assessment and the new standards project: A story of serendipitous success. *Phi Delta Kappan*, 8, 758-764.

About the Authors

Dorothy S. Strickland is the Samuel DeWitt Proctor Professor of Education at Rutgers University. A former classroom teacher, reading consultant, and learning disabilities specialist, she is a past president of both the International Reading Association and the IRA Reading Hall of Fame. She received IRA's Outstanding Teacher Educator of Reading Award. She was the 1998 recipient of the National Council of Teachers of English Award as Outstanding Educator in the Language Arts and the 1994 NCTE Rewey Belle Inglis Award as Outstanding Woman in the Teaching of English. Recent publications include: *Beginning Reading and Writing, Administration and Supervision of Reading Programs,* and *Supporting Struggling Readers and Writers.*

Catherine E. Snow is the Henry Lee Shattuck Professor of Education at the Harvard Graduate School of Education. She received a Ph.D. in psychology from McGill University, then worked for several years in the linguistics department of the University of Amsterdam, focusing on first language development in young children and on second language acquisition. After moving to HGSE, her interests expanded to include the language learning of school aged children and literacy development in both first and second languages. She is a past president of AERA and a member of the National Academy of Education. She chaired the NRC committee that produced *Preventing Reading Difficulties in Young Children,* and the RAND Reading Study Group that produced *Reading for Understanding: Towards an R&D Agenda in Reading.*

Peg Griffin studies language as a medium and topic of instruction, usually in collaboration with other linguists and psychologists. Her Ph. D. is from Georgetown University and she has learned from teaching at all the levels of the education system. She worked with the Committee on the Prevention of Reading Difficulties in Young Children and continues to worry about policy and teacher preparation regarding early reading. She studies learning in preschool beyond language and literacy, revisiting earlier interest in math and science. She has renewed her research affiliation with the Laboratory of Comparative Human Cognition at the University of California at San Diego.

M. Susan Burns is currently a faculty member at George Mason University. She held faculty appointments at Tulane University and the University of Pittsburgh. At the National Research Council, she was study director for the Committee on the Prevention of Reading Difficulties in Young Children. Her research interests include instructional practices that facilitate early language and literacy development, assessment of young children, early childhood curriculum, and parent-child interaction. Applied interests include the development of intervention/prevention strategies for children living in poverty, those who have a home language other than English, and children with disabilities. She has a Ph.D. in applied developmental psychology from Peabody College, Vanderbilt University.

Peggy McNamara is the Director of the Reading and Literacy Program at Bank Street College of Education. She also supervises teachers in field placements and teaches a range of literacy courses that prepare teachers to work with students who have diverse learning needs. Through an OERI three-year grant, she studied First Steps, a literacy resource which had been developed in Western Australia. She examined the effectiveness of First Steps within the context of an urban New England school district and the implication of its outcomes for literacy education. An educator for the past twenty-six years, Peggy has worked in public and private schools in New York City serving as a classroom teacher and professional developer.

Index

Credits

Page i, ThinkStock; **page v,** (top) courtesy of Jim Harris, Children's House Montessori School, Burlington, North Carolina; (bottom) BrandX Pictures; **page vi,** (top) PhotoDisc; (bottom) Peggy McNamara, Reading and Literacy Program, Bank Street College of Education; **page vii,** (top) Jim Harris; (bottom) Banana Stock; **page 1,** (top left) Rebecca Brittain, Rutgers University, Graduate School of Education, 2001 Summer Institute, "Helping Teachers Teach Reading"; (top right) Banana Stock; **page 2,** Jim Harris; **page 5,** (top) San Diego Unified School District; (bottom) Peggy McNamara; **page 8,** PhotoDisc; **page 10,** Jim Harris; **page 19,** StockByte; **page 20,** ThinkStock; **page 29,** (top left) Rebecca Brittain; (top right) Banana Stock; **page 30,** Lifetouch National School Studios Inc.; **page 32,** Jennifer L. Dewitt, George Mason University, Graduate School of Education; **page 33,** Jennifer L. Dewitt; **page 36,** Jim Harris; **page 39,** BrandX Pictures; **page 55,** (top left) Rebecca Brittain; (top right) Paul Hartmann, Photography; **page 59,** ThinkStock; **page 65,** Josh Gottesman, Larchmont, New York; **page 67,** Digital Stock; **page 68,** StockByte; **page 69,** PhotoDisc; **page 81,** (top left) Rebecca Brittain; (top right) Photodisc; **page 84,** BrandX Pictures; **page 88,** Jim Harris; **page 89,** Jim Harris; **page 93,** Lawrence Nathan; **page 94,** courtesy Tools of the Mind Research Project, D. J. Leong and E. Bodrova, leongd@mindspring.com; **page 95,** Tools of the Mind Research Project; **page 96,** Tools of the Mind Research Project; **page 110,** ThinkStock; **page 119,** (top left) Rebecca Brittain; (top right) Photodisc; **page 125,** Chelsey E. Goddard; **page 126,** Digital Stock; **page 127,** Jim Harris; **page 128,** Digital Stock; **page 130,** courtesy Evan Greenleaf King, Albany, California; **page 132,** (For more information on this instrument and research using it, see McKenna & Kear, 1990.); **page 147,** (top left) Rebecca Brittain; (top right) Photodisc; **page 152,** BrandX Pictures; **page 154,** Banana Stock; **page 156,** Sharon Back; **page 159,** Tools of the Mind Research Project.